CARL G. RASMUSSEN

ESSENTIAL ATLAS OF THE BIBLE

ZONDERVAN

OF THE

ZONDERVAN ACADEMIC

ZONDERVAN ACADEMIC

Zondervan Essential Atlas of the Bible
Copyright © 2013 by Carl G. Rasmussen

Published in Grand Rapids, Michigan, by Zondervan. Zondervan is a registered trademark of The Zondervan Corporation, L.L.C., a wholly owned subsidiary of HarperCollins Christian Publishing, Inc.

Requests for information should be addressed to customercare@harpercollins.com.

Zondervan titles may be purchased in bulk for educational, business, fundraising, or sales promotional use. For information, please email SpecialMarkets@Zondervan.com.

ISBN 978-0-310-51811-2 (ebook)

Library of Congress Cataloging-in-Publication Data

Rasmussen, Carl.
     Zondervans essential atlas of the Bible / Carl G. Rasmussen.
          p. cm.
     Includes bibliographical references and index.
     ISBN 978-0-310-31857-6 (pbk.)
     1. Bible — Geography — Maps. 2. Bible — History of Biblical events — Maps. I. Title.
G2230.R33 2014
220.9'10223 — dc23   2012589773

Cover design: www.wdesigncompany.com
Cover photography: Ben Greenhoe/Zondervan; Michael Melford/Getty Images;
     www.holylandphotos.org; Slow Images/Getty Images
Interior design: Kirk DouPonce

Printed in the United States of America

24 25 26 27 28 29 30 31 32 33 34 /TRM/ 23 22 21 20 19 18 17 16 15 14 13 12

CONTENTS

PREFACE AND ACKNOWLEDGMENTS

This volume is an adaption of the more complete *Zondervan Atlas of the Bible*. It begins with a concise Geographical Section, which introduces to the reader the lands of the Bible, including Israel/Jordan, Egypt, Syria and Lebanon, and Mesopotamia. Helpful maps, charts, and pictures illustrate the topic discussed — topography, regions, weather, roads, and so on.

The Geographical Section is followed by a Historical Section, which provide maps, commentary, diagrams, and pictures related to the whole sweep of biblical history, both Old and New Testaments — including the Intertestamental Period. Jerusalem and the seven churches of Revelation have chapters of their own. Readers who are interested in fuller discussions of specific places, regions, and events are invited to consult the *Zondervan Atlas of the Bible*, *The Zondervan Encyclopedia of the Bible*, and the author's website, www.HolyLandPhotos.org.

This atlas is intended for readers of the Bible who want concise information close by as they are reading the biblical text. It is ideal for use by Bible study groups, adult Bible classes, and travelers to the Middle East, and it will serve well as an auxiliary textbook for college, university, and seminary classes.

I wish to thank acquisitions editor David Frees, who helped initiate this project, and acquisitions editor Madison Trammel, who has seen it through to completion. As with the *Zondervan Atlas of the Bible*, Kim Tanner's valuable expertise shows through in the development and presentation of maps, graphics, and pictures. Mark Connally has graciously allowed the use of some of his beautiful images in this volume. Mark Sheeres and his team have developed a layout that has made the volume a pleasure to view and use. I was greatly involved in revising the text of the larger work for this volume, but I would especially like to thank Verlyn Verbrugge, whose wisdom and skill brought this process to a successful conclusion. In addition, I would like to thank Stanley Gundry and Paul Engle for their encouraging support through the years.

Finally I would like to thank my wife, Mary, whose love, companionship, and encouragement over the years has been a blessing to me and to all those with whom she comes in contact.

Carl G. Rasmussen

ABBREVIATIONS

AD	*Anno Domini*, In the year of our Lord (i.e., after the birth of Christ)	Jos.*Apion*	Josephus: *Against Apion*
ANET	J. B. Pritchard, ed., *Ancient Near Eastern Texts Relating to the Old Testament*. Third edition. Princeton: Princeton University Press, 1969.	Jos.*Life*	Josephus: *Life*
		Jos.*War*	Josephus: *The Jewish War*
		Kh.	Khirbet
		LB	Late Bronze Age
		MB	Middle Bronze Age
BC	Before Christ	mi.	mile(s)
c., ca.	*circa*, about	Mt(s).	Mountain(s)
ch., chs.	chapter(s)	N.	Nahr/Nahal
e.g.	for example	NASB	New American Standard Bible
EB	Early Bronze Age	NIV	New International Version
ed., eds.	editor(s)	NT	New Testament
esp.	especially	OT	Old Testament
et al.	and others	p., pp.	page(s)
etc.	*et cetera*, and so on	par., pars.	paragraph(s)
F	Fahrenheit	R.	River
ft.	foot, feet	sq.	square
H.	Horbat	T.	Tell (Arabic)/Tel (Hebrew)
Heb.	Hebrew	v., vv.	verse(s)
in.	inch(es)	W.	Wadi
Jos.*Antiq.*	Josephus: *The Antiquities of the Jews*		

GEOGRAPHICAL
SECTION

INTRODUCTION TO THE MIDDLE EAST AS A WHOLE

The stage on which the major events of Old Testament history took place includes all the major countries shown on page 9. This large land mass is bounded on the west by the Nile River and the Mediterranean Sea, on the north by the Amanus and Ararat Mountains, and on the east by the Zagros Mountains and the Persian Gulf. To the south, the Nafud Desert and the southern tip of Sinai form a rather loose boundary.

Much of the Middle East is desert. Large portions of modern-day Syria, Iraq, Jordan, and Saudi Arabia include desert wastes such as the Syrian Desert, the Nafud, the Arabian Desert, the Ruba al-Khali, Negev, Sinai, and Egypt. The seas and gulfs that help outline the Middle East have influenced life in the area. The most important of these is the Mediterranean Sea, which offers life-giving rains to most of the region. Much of what has occurred in the Middle East can be summed up as a struggle between the influences of the desert and the Mediterranean Sea over against the people who have lived there.

The first section of this book outlines briefly some of the significant challenges of this part of the world — geography, climate, roads, trade routes, food supply, and the like. It is easy to determine where the majority of people have lived in the Near East by highlighting on a map (see p. 9) the areas watered by the Nile, the Tigris, and the Euphrates, as well as those regions that receive over twelve inches of rainfall annually. This area is roughly the shape of a crescent, with one point in the Nile River and the other in the Persian Gulf. It is aptly named the "Fertile Crescent."

▼ *Roman road in Syria*

Black Sea

MACEDONIA
Skopje **BULGARIA**
Tirana
ALBANIA

GEORGIA Tbilisi **RUSSIA**
ARMENIA **AZERBAIJAN**
Yerevan
Ararat Mt. Baku

GREECE

Ankara

TURKEY

Halys R.

Caspian Sea

Athens

Aegean Sea

Taurus Mts.

Amanus Mts.

Euphrates R.

Tigris R.

Zagros Mts.

IRAN

Tehran

Mediterranean Sea

Nicosia
CYPRUS
Beirut
LEBANON
Golan Heights
ISRAEL
Jerusalem

SYRIA
Damascus

Baghdad

IRAQ

Jordan R.

Amman

JORDAN

Gaza Strip
West Bank
Cairo

Sinai

An Nafud Desert

KUWAIT
Kuwait

Persian Gulf

LIBYA

EGYPT

Nile R.

Red Sea

SAUDI ARABIA

Arabian Peninsula

BAHRAIN
Manama

Doha
QATAR

Riyadh

0 — 200 km.
0 — 200 miles

Black Sea
Sinop
Amisus

EUROPE

Delphi
Mycenae
Pylos Sparta Athens
MYCENAEAN
MINOAN
Knossos
Crete Phaistos

Miletus

Rhodes

Attalia

Hattusa

HITTITE - HURRIAN

Kanish

Tarsus

Cyprus

Carchemish

Haran

Tigris R.

Nineveh

ASIA

Caspian Sea

Ugarit
Hamath
Arvad
Byblos
Sidon
Tyre

Ebla

Aleppo

Gozan

ASSYRIA

Asshur

Nuzi

Ecbatana

Mediterranean Sea

Qatna

AMORITE

Euphrates R.

Tadmor
Mari

Akkad?

Susa

Damascus
Hazor
Megiddo
Ramoth Gilead

Babylon

BABYLONIA

Cyrene

Alexandria

Zoan

Gaza
Jerusalem

Ur

AFRICA

On
Memphis

EGYPTIAN

Kadesh Barnea

Elath

Dumah

Arabian Desert

Persian Gulf

Tema

Nile R.

Red Sea

Thebes

0 — 200 km.
0 — 200 miles

	Fertile Crescent
	Land routes
	Sea routes
AMORITE	Cultural spheres

THE GEOGRAPHY OF ISRAEL AND JORDAN

Terrain

At the southeastern end of the Mediterranean Sea, we can distinguish five major longitudinal zones. As one moves from west to east they are: the coastal plain, the central mountain range, the rift valley, the Transjordanian mountains, and the eastern desert.

(1) The **coastal plain** stretches approximately 120 miles along the Mediterranean coast from Rosh HaNiqra south to Gaza. It receives 25 to 16 inches of rain per year, the northern sections receiving more rain than the southern. A few powerful springs provided water, but more commonly the

▼ *Jezreel Valley from Megiddo looking east at Mount Tabor*

inhabitants used wells to tap the water table. The coastal plain consists mainly of low, rolling hills covered with fertile alluvial soils. Grain crops flourished in the winter and spring months, while flocks grazed there during the remainder of the year.

While travel was easy in this area, travelers did have to be careful to avoid sand dunes, large rivers such as the Yarkon River, and low-lying areas that became swampy during the winter months. Also, they had to choose the most appropriate track through Mount Carmel. The only natural seaport is at Acco.

(2) **The central mountain range** runs from Galilee in the north to the Negev Highlands in the south. It rises in places to more than 3,000 feet and is severed in an east–west direction by the Jezreel Valley in the north and the Negev Basin in the south, where east–west traffic can flow with relative ease.

Cutting through the limestone hills are deep V-shaped valleys, usually called wadis. They are dry during the summer months but sometimes flow with water during the winter. They drain either toward the rift valley or the Mediterranean Sea. Travel along the bottoms of these deep wadis is difficult because of boulders and occasional cliffs, and north–south travel across the wadis is almost impossible. Thus roads tended to be located on the mountain ridges.

The western slopes of the mountains receive considerable rainfall (20 to 40 in.); this, along with the fertile soil, ensures the fertility of the area. Here — largely on hillside terraces partially formed by the natural bedding of the limestone — small fields of wheat, groves of olive trees, and vineyards flourish (Deut 8:8; Ps 147:14; Hab 3:17 – 19).

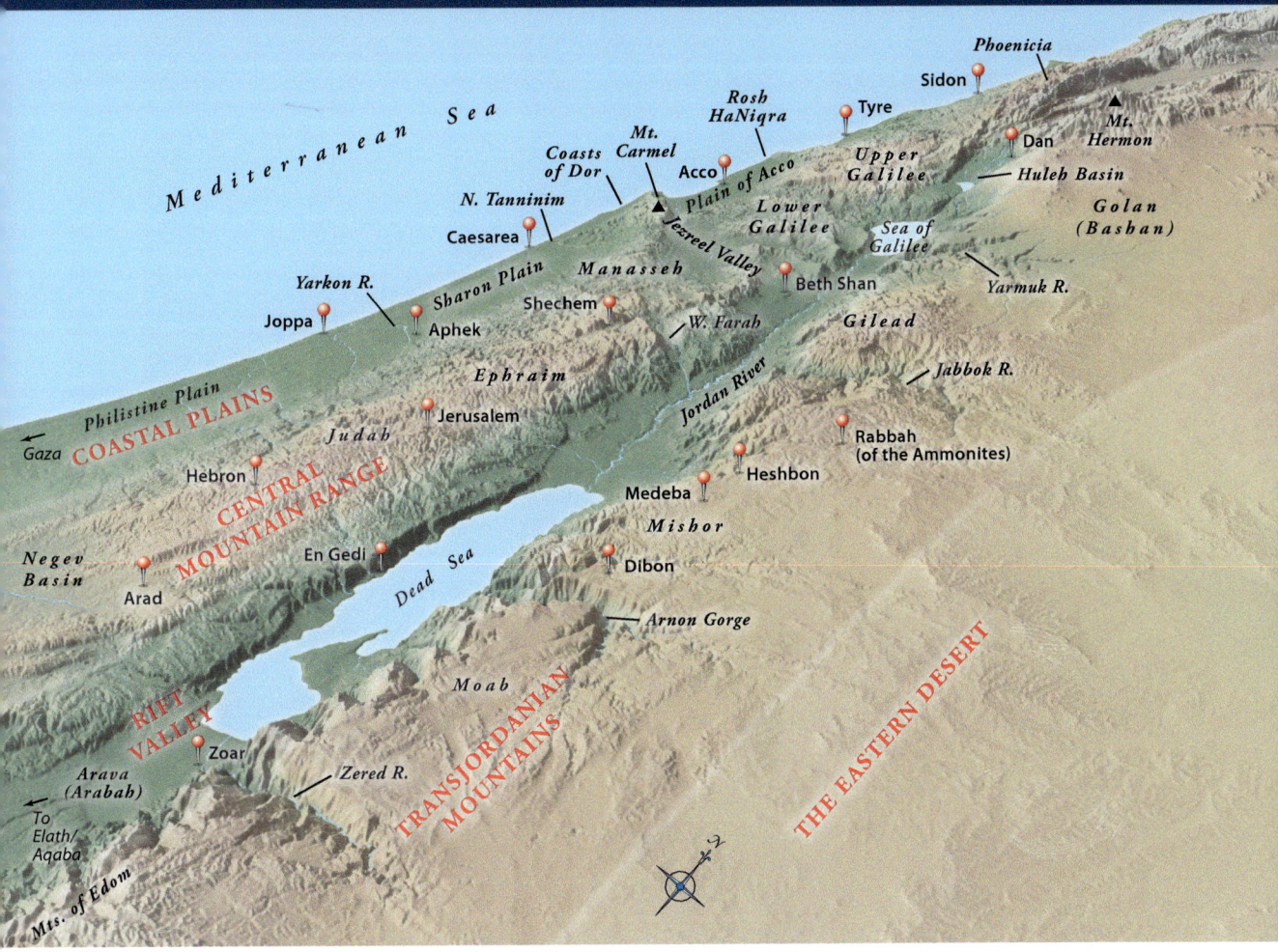

Winter rainwater seeps into the limestone until it reaches an impermeable layer, where it begins to flow laterally until it emerges as a spring. Settlements often developed close to these freshwater springs, but being on the slopes of the hills they were difficult to defend. By about 1400 BC, the construction of cisterns, lined with plaster to prevent leakage, began to solve the problem of complete dependency on natural water sources.

The Israelites first settled in the central mountain range. Because international powers were primarily interested in controlling the coastal plain, the mountains provided the Israelites with security. Only during periods when they considered their power to be great did the Israelites attempt to gain control of the coastal plain, but this almost always resulted in conflict with one or more of the great powers.

(3) The next zone, part of **the rift valley system** that continues into Africa, stretches 260 miles from Dan to Elath at the northern tip of the Red Sea. A considerable amount of rain falls in the northern section of this zone (24 in. at Dan), whereas in the south rainfall is negligible (2 in. at the south end of the Dead Sea).

The northernmost section of the rift valley, called the Huleh Basin, receives about 24 inches of rain each year. Springs at the foot of Mount Hermon form the headwaters of the Jordan River and flow through a marshy lake known in antiquity as Lake Semechonitis. The Jordan then enters the north end of the Sea of Galilee, which lies 690 feet below sea level and measures 13 by 7.5 miles. The temperate Mediterranean climate makes this region a desirable place to live. The sea itself is a major source of fish for inhabitants, and a number of small but fertile plains along the sea's shoreline have been intensively cultivated throughout history.

The Jordan River flows out of the Sea of Galilee and descends to the Dead Sea. The linear distance in the Jordan Valley is 65 miles, but the length of the river as it winds its way is 135 miles. Until modern times, when Israelis and Jordanians began diverting water for commercial purposes, the Jordan averaged 100 feet in width with a depth of 3 to 10 feet. After heavy rains in late winter and spring its width could swell to almost a mile in places.

The Jordan River empties into the Dead Sea — the lowest spot on the surface of the earth (1,385 ft. below sea level). This sea does not have any outlet and is called the "Salt Sea" because of its high mineral content. South of the Dead Sea, the rift valley continues 110 miles to the shores of the Red Sea. This region is called the "Arava" or "Arabah" on modern Israeli maps, although the biblical Arabah was primarily north of the Dead Sea (e.g., Deut 3:17; Josh 11:2; 2 Sam 2:29). Elath marks the southern boundary of modern Israel and, at times, of biblical Israel.

(4) Next are the **mountains of Transjordan**, stretching from Mount Hermon in the north to the Gulf of Aqaba/Elath in the south on the east side of the Jordan. While the western slopes of these mountains are often steep, the eastern slopes descend gradually into the eastern desert.

Some of the biblically recognizable places, from north to south, are: the region of Bashan, the region of Gilead (with the Yarmuk and the Jabbok Rivers), and Moab (between the Arnon and Zered Rivers). The

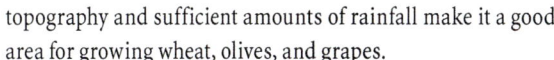
▲ *River Jordan with surrounding "thicket" (= Zor)*

▲ *Wadi with water in the Judean Wilderness. Notice the barren limestone slopes.*

topography and sufficient amounts of rainfall make it a good area for growing wheat, olives, and grapes.

South of the Zered Valley are the mountains of Edom, extending to Aqaba. Along the western crest of this ridge there is sufficient rainfall for growing wheat and barley. The most famous city of this remote region is Petra. The major road east of the rift valley was the Transjordanian Highway that connected Damascus with the countries located in present-day Saudi Arabia. The southern portion of the high-

way, near Heshbon, was called the "King's Highway" (Num 21:22), although this name was used for another road as well (Num 20:17).

(5) Finally is **the eastern desert** is located to the east of the Transjordanian Mountains. In the north, the great volcanic mountains and lava make the region inhospitable, but its high elevation ensures adequate rainfall to grow crops. The barren desert stretches eastward some 400 miles to the Euphrates River.

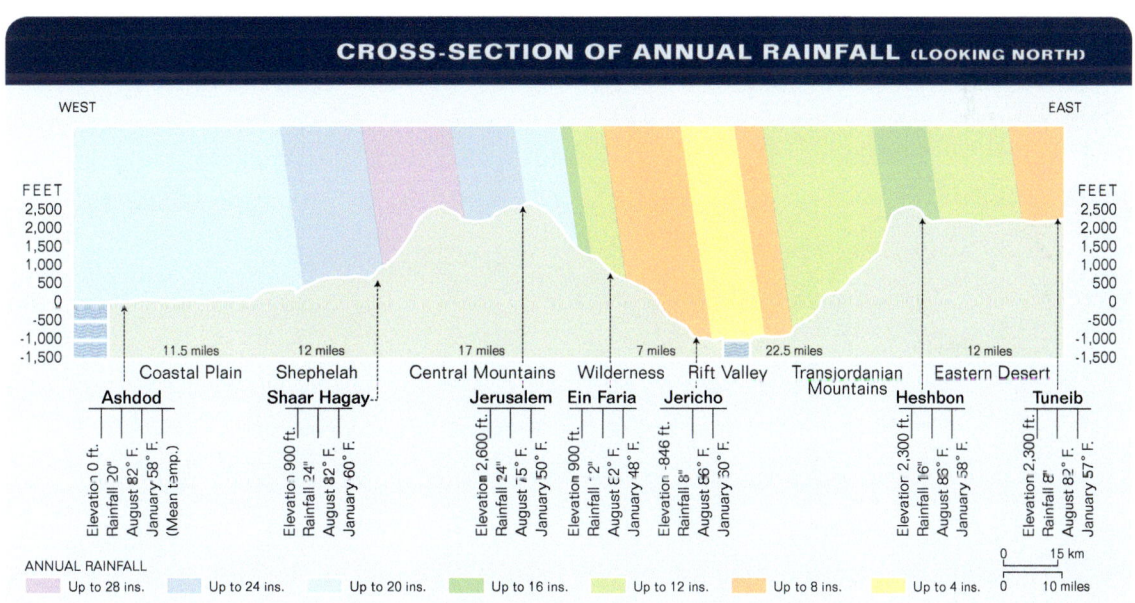

CROSS-SECTION OF ANNUAL RAINFALL (LOOKING NORTH)

WEST   EAST

FEET	FEET
2,500	2,500
2,000	2,000
1,500	1,500
1,000	1,000
500	500
0	0
-500	-500
-1,000	-1,000
-1,500	-1,500

11.5 miles   12 miles   17 miles   7 miles   22.5 miles   12 miles

Coastal Plain   Shephelah   Central Mountains   Wilderness   Rift Valley   Transjordanian Mountains   Eastern Desert

Ashdod   **Shaar Hagay**   **Jerusalem**   **Ein Faria**   **Jericho**   **Heshbon**   **Tuneib**

Ashdod: Elevation 0 ft. / Rainfall 20" / August 82° F. / January 58° F. (Mean temp.)

Shaar Hagay: Elevation 900 ft. / Rainfall 24" / August 82° F. / January 60° F.

Jerusalem: Elevation 2,600 ft. / Rainfall 24" / August 75° F. / January 50° F.

Ein Faria: Elevation 900 ft. / Rainfall -2" / August 82° F. / January 48° F.

Jericho: Elevation -846 ft. / Rainfall 8" / August 86° F. / January 30° F.

Heshbon: Elevation 2,300 ft. / Rainfall 16" / August 86° F. / January 58° F.

Tuneib: Elevation 2,300 ft. / Rainfall 8" / August 82° F. / January 57° F.

ANNUAL RAINFALL

Up to 28 ins.	Up to 24 ins.	Up to 20 ins.	Up to 16 ins.	Up to 12 ins.	Up to 8 ins.	Up to 4 ins.

0   15 km
0   10 miles

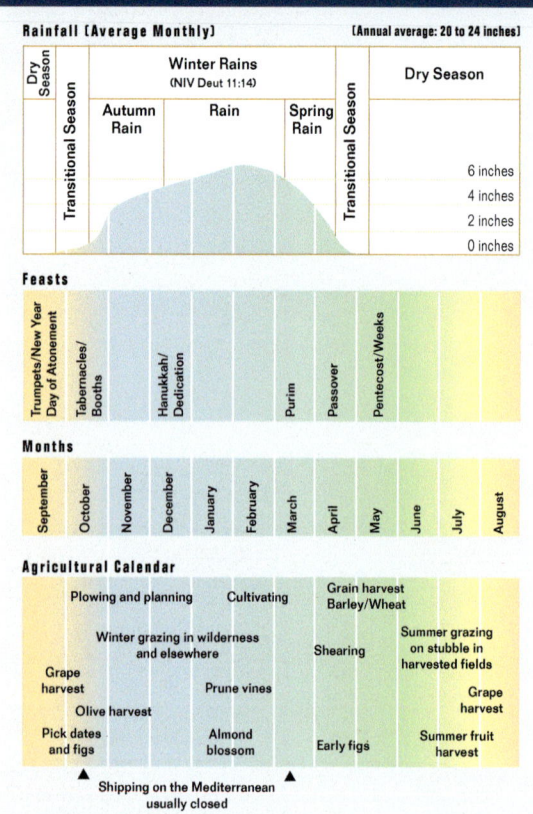

RAINFALL, AGRICULTURAL, AND PASTORAL PATTERNS IN THE JERUSALEM AREA

Rainfall (Average Monthly) — [Annual average: 20 to 24 inches]

Dry Season | Transitional Season | Winter Rains (NIV Deut 11:14) — Autumn Rain / Rain / Spring Rain | Transitional Season | Dry Season

6 inches
4 inches
2 inches
0 inches

Feasts

Trumpets/New Year / Day of Atonement · Tabernacles/Booths · Hanukkah/Dedication · Purim · Passover · Pentecost/Weeks

Months

September · October · November · December · January · February · March · April · May · June · July · August

Agricultural Calendar

Plowing and planning · Cultivating · Grain harvest Barley/Wheat

Winter grazing in wilderness and elsewhere · Shearing · Summer grazing on stubble in harvested fields

Grape harvest

Olive harvest · Prune vines · Grape harvest

Pick dates and figs · Almond blossom · Early figs · Summer fruit harvest

Shipping on the Mediterranean usually closed

Climate

Israel's year is divided into two major seasons: the rainy season (mid-October through April) and the dry season (mid-June through mid-September). Climactic conditions in Israel during the summer months are relatively stable. Warm days and cooler nights are the rule, and it almost never rains. In Jerusalem, for example, the average August daytime high is 86°F, the nighttime average low is 64°F.

During the summer, olives, grapes, figs, pomegranates, melons, and other crops are ripening and being tended by farmers. Most fruits are harvested in August and September. During the summer, shepherds move their flocks of sheep and goats westward, allowing them to feed on the stubble of wheat and barley fields that were harvested in the late spring. Because the soil is dry during summer months, travel is easy, and caravans and armies moved through most parts of the country without difficulty; the armies often helped themselves to the plentiful supplies of grain at the expense of the local populace.

The rainy season is much cooler. During January the mean daily temperature in Jerusalem is 50°F, and the city receives snow once or twice each year. Life is uncomfortable in the hilly regions — a discomfort the people gladly bear because of the life-giving power of the rains. The Bible actually refers to three parts of the rainy season in Deuteronomy 11:14: "Then I will send rain [Heb. *matar*; Dec.–Feb.] on your land in its season, both autumn [Heb. *yoreh*; Oct.–Dec.]

▼ *Wheat harvest in early summer*

and spring rains [Heb. *malqosh*; March – April], so that you may gather in your grain, new wine and olive oil" (cf. also Jer 5:24; Hos 6:3). Note the following:

- The amount of rainfall decreases as one moves from north to south.
- The amount of rainfall decreases as one moves from west to east, away from the Mediterranean Sea.
- The amount of rainfall increases with the elevation.
- The amount of rainfall is greater on the windward (Mediterranean) side of the mountains than on the leeward side.

During a typical year a farmer plows his field and plants his grain crops after the "autumn rains" of October through December have softened the hard, sunbaked soil. The grain crops ripen during March and April, as the rains begin to taper off. These "spring rains" are important for producing bumper crops.

There are two transitional seasons. One lasts from early May through mid-June. It is punctuated by a series of hot, dry, dusty days — which are called by the names of these winds: hamsin or sirocco. Hamsin conditions can sap the energy of both humans and animals, and they completely dry up the beautiful flowers and grasses that cover the landscape during the winter months (Isa 40:7 – 8). But these same hot, dry winds aid the ripening of grains by "setting" them before the harvest.

The second transitional season, from mid-September to mid-October, marks the end of the stable, dry, summer conditions. It is the time of the fruit harvest, and farmers begin to look anxiously for the onset of the rainy season. In the fall, travel on the Mediterranean becomes dangerous (Acts 27:9), and it remains so throughout the winter months.

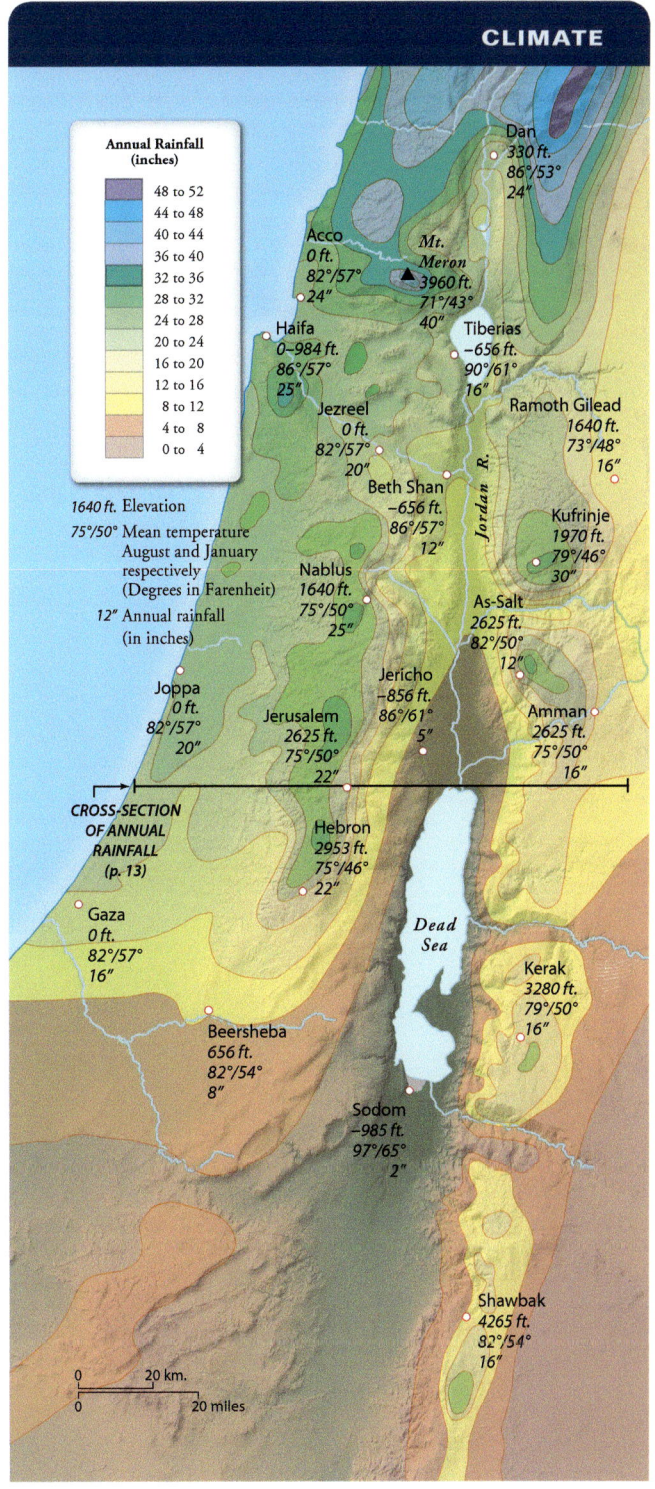

CLIMATE

Annual Rainfall (inches)

- 48 to 52
- 44 to 48
- 40 to 44
- 36 to 40
- 32 to 36
- 28 to 32
- 24 to 28
- 20 to 24
- 16 to 20
- 12 to 16
- 8 to 12
- 4 to 8
- 0 to 4

1640 ft. Elevation
75°/50° Mean temperature August and January respectively (Degrees in Farenheit)
12" Annual rainfall (in inches)

CROSS-SECTION OF ANNUAL RAINFALL (p. 13)

Dan 330 ft. 86°/53° 24"

Acco 0 ft. 82°/57° 24"

Mt. Meron 3960 ft. 71°/43° 40"

Haifa 0–984 ft. 86°/57° 25"

Tiberias –656 ft. 90°/61° 16"

Jezreel 0 ft. 82°/57° 20"

Ramoth Gilead 1640 ft. 73°/48° 16"

Beth Shan –656 ft. 86°/57° 12"

Kufrinje 1970 ft. 79°/46° 30"

Nablus 1640 ft. 75°/50° 25"

As-Salt 2625 ft. 82°/50° 12"

Joppa 0 ft. 82°/57° 20"

Jerusalem 2625 ft. 75°/50° 22"

Jericho –856 ft. 86°/61° 5"

Amman 2625 ft. 75°/50° 16"

Jordan R.

Hebron 2953 ft. 75°/46° 22"

Gaza 0 ft. 82°/57° 16"

Dead Sea

Kerak 3280 ft. 79°/50° 16"

Beersheba 656 ft. 82°/54° 8"

Sodom –985 ft. 97°/65° 2"

Shawbak 4265 ft. 82°/54° 16"

0   20 km.
0   20 miles

Map labels:

Mediterranean Sea
Sidon
Litani R.
Beqa
Mt. Hermon
To Tadmor
Damascus
Tyre
Janoah
Dan
Abel Beth Maacah
Hazor
Acco
Arbel Cliffs
Sea of Galilee
N. Raqqad
Karnaim
Hannathon
Ashtaroth
Mt. Carmel
Jokneam
▲Mt. Tabor
Yarmuk R.
Megiddo
Aruna
Beth Shan
Ramoth Gilead
International North–South Route
Dothan
Yaham
Jordan R.
Samaria
Tirzah
Jabbok R.
Shechem
N. Yarkon
Jokmeam
Aphek
Adam
Joppa
Lebonah
Shiloh
Ramah
Bethel
Rabbah (of the Ammonites)
Upper Beth Horon
Jericho
Ashdod
Gezer
Micmash
Heshbon
Beth Shemesh
Jerusalem
Gath
Gaza
Hebron
Dead Sea
Aroer
Yurza
Gerar
En Gedi
Arnon R.
Raphia
N. Besor
Beersheba
Arad
Elusa
Negev
Mampsis
Kir Moab
Zoar
Zered R.
N. Zin
Avedat
Tamar
Bozrah
Kadesh Barnea
Transjordanian Route
International Route
Petra
To Egypt
Elath?

Legend:
— International routes
— Interregional and local routes

0   20 km.
0   20 miles

Roads and Travel

The roads that developed in ancient Israel can be divided into three major categories: international, interregional, and local. The international and interregional roads were for commercial purposes — for transporting items such as foodstuffs, cloth, metals, incense, and fine pottery. These roads also served as thoroughfares for military expeditions and itinerant tradesmen, for the migration of peoples, for the conveyance of governmental and commercial messages, and for the travel of pilgrims to holy places.

Those who controlled the roads could collect tolls from passing caravans, sell food and lodging, and "offer" the services of military escorts to "ensure" travelers' safe passage through "dangerous" territory. Those living along the international routes were exposed to new intellectual, cultural, lin-

▲ *Jerusalem during a hamsin*
▼ *Jerusalem three days after the hamsin*

▲ *Yarmuk River in the southern Bashan*

guistic, and religious influences, but they were also exposed to the ravages of war as armies moved along these same paths.

Besides walking, early modes of transportation included donkeys, carts, and chariots and horses. Camels were domesticated to carry heavy loads. People preferred to travel during the dry summer season rather than attempt to negotiate muddy, rain-soaked terrain in the winter months. In the spring was "the time when kings go off to war" (2 Sam 11:1) because the roads were dry and harvested grain was available to feed their troops.

The most important *international* route through Israel connected Egypt with its rivals/allies to the north and east (Hittites, Hurrians, Syrians, Assyrians, Babylonians, Persians, etc.). This international route is sometimes incorrectly called the "Way of the Sea" (cf. Isa 9:1) or the "Via Maris." At Yaham a traveler chose one of several passes that led through Mount Carmel. Several options were available for travel from Megiddo to Damascus, from where one could proceed to Turkey or the Euphrates River.

The other international route led south from Damascus and traveled the entire length of Transjordan. One branch of this route ran just east of the Transjordanian Mountains, where there were good supplies of water, but a traveler also had to cross difficult wadis, such as the Yarmuk. The other branch ran further east along the edge of the desert, where not as much water was available, and caravans traveling along it were subject to raids by desert tribes.

The *interregional* route that ran from Beersheba in the south to Shechem in the north — via Hebron, Bethlehem, Jerusalem, Ramah, Bethel/Ai, and Shiloh — is sometimes called the "Route of the Patriarchs" because Abraham, Isaac, and Jacob traveled its length. Others refer to it as the "Ridge Route," for in many places it "tiptoes" along the watershed of the Judean and Ephraim mountains. This road furnishes the backdrop for many events recorded in the Bible.

THE GEOGRAPHY OF EGYPT

Terrain

Located in the northeastern corner of Africa, Egypt has been one of the great power centers of the Near East. Its heartland is basically a long river oasis situated near the eastern edge of the Sahara Desert. Ninety-five percent of Egypt is stone, sand, and desert, while only 5 percent is rich agricultural land, to which the life-giving Nile brings precious water and silt. The northward-flowing Nile, with its origins in Central Africa, is the longest river in the world (4,145 mi.).

The traditional boundaries of ancient Egypt were the Mediterranean Sea on the north, the Red Sea/Gulf of Suez on the east, the first cataract (= rapids) of the Nile near Aswan on the south, and a north–south line of oases about 120 miles west of the Nile.

Egypt was divided into two major geographical regions. "Upper Egypt," which is upstream (i.e., south), stretches from the first cataract in the south to the beginning of the delta near Cairo, while "Lower Egypt" is the delta proper. In Upper Egypt the arable land lies along both sides of the Nile.

In ancient times, the Nile usually rose between 15 and 23 feet, overflowed its banks, and flooded the nearby fields. The muddy floodwaters covered the fields for several months; as they began to recede in September/October, they leeched out unwanted salts and left behind a fresh layer of fertile silt. The peasants planted wheat and barley in the muddy soil during October/November and harvested them from January through March. They also ate fish from the river and fowl, such as ducks and geese. Flax was grown for clothing, ropes, and

sails, while papyrus was used for paper production and was exported.

The Nile itself was the major "road," and barges and sailing ships were common modes of transportation. The current easily carried these vessels downstream, and the Egyptians were able to sail upstream by making use of the prevailing north wind.

In ancient times seven branches of the Nile in Lower Egypt made their way through the delta area to the Mediterranean Sea. This low-lying area also was well supplied with fertile silt washed down over the millennia and was crisscrossed by

YEARLY CYCLE IN ANCIENT EGYPT

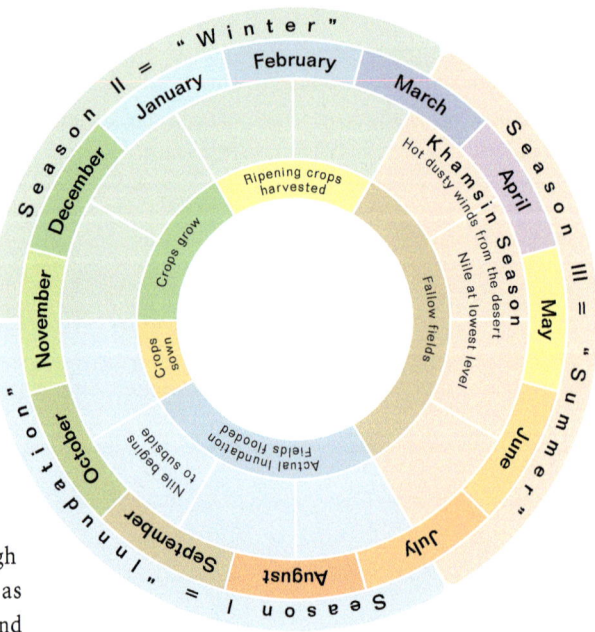

Mediterranean Sea

Damascus

International routes
Local or regional routes
Water routes

Tel Aviv
Amman
Gaza
Raphia
Jerusalem
Alexandria
Port Said
El-Arish
Beersheba

LOWER EGYPT
Delta
Ismailiya
Wadi el-Arish
GOSHEN
Cairo
Suez
Giza
Memphis
Jebel et Tih
Timnah
Saqqara
Lisht
Elath
Faiyum Oasis
Aqaba

Bahariya Oasis
Gulf of Suez
Sinai Peninsula
Gulf of Elath/Aqaba
El Amarna
MIDIAN

Farafra Oasis
UPPER EGYPT
Sharm esh-Sheikh
ARABIA

Nile R.
Eastern Desert
Abydos

Ed-Dakhla Oasis
Thebes
Wadi Hammamat

Ed-Kharga Oasis
Jebel es-Silsila
Medina
Dunqul Oasis
Kurkur Oasis
Aswan
1st Cataract
Berenike

Salima Oasis
2nd Cataract
Red Sea

NUBIA
Mecca

3rd Cataract
Jebel Barkal
Nile R.
4th Cataract
5th Cataract

Sahara Desert
CUSH

6th Cataract
Atbara R.

White Nile R.
Blue Nile R.
ETHIOPIA
Bab al Mandab

Ethiopian Highlands

0   100 km.
0   100 miles

▲ *Canal near the Nile River bringing water to the fields*

canals used for irrigation and transport. In addition to crops, swamps provided some pasturage for cattle.

From the eastern delta the armies of the mighty pharaohs of the Eighteenth and Nineteenth Dynasties (ca. 1500 – 1150 BC) launched their expeditions into Canaan and other countries of Asia. According to Genesis, it was in the rich and fertile eastern delta, known as the "region of Goshen," that Jacob and his descendants settled and began their sojourn in Egypt (Gen 46 – 50).

To the east of the delta lies the triangular Sinai peninsula. Its northern boundary is the Mediterranean Sea; this coastline is primarily sandy flats and some dunes. The major road connecting Asia with Africa ran through this area. Not only did commercial caravans use it, but the great armies of the world have passed this way. Most of this region is drained by the Wadi el-Arish ("Brook of Egypt"), which enters the Mediterranean Sea at el-Arish.

SINAI

Mediterranean Sea

Gaza
Salt Sea
Arad
Beersheba
El-Arish
Kadesh Barnea
W. el-Arish
Petra
E D O M
To Arabia →
"Arabah"
Way to Shur
Darb el-Gaza
Timnah
Elath
Aqaba
A R A B I A
Lake Timsah
Bitter Lakes
Darb el-Hagg
Tell er-Retabah
Goshen
Suez
Bir Mara
Et-Tih Desert
Ayun Musa
Jebel Sin Bisher
M I D I A N
E G Y P T
Ein Hawwara
W. Gharandal
Jebel et-Tih
Serabit el-Khadim
W. Feiran
Gulf of Aqaba
Feiran Oasis
St. Catherine's Monastery
Jebel Sirbal
Jebel Musa
Gulf of Suez
N
Sharm esh-Sheikh
Red Sea

▲ *Luxor: Looking west across the Nile at agriculture and the mountains by the Valley of the Kings* Mark Connally

To the south, the dunes eventually give way to a series of mountains. Water supplies are found in and near these mountains, particularly in the northeast at Kadesh Barnea, where the most powerful spring of the peninsula is located. The southern tip of Sinai consists of dramatic, jagged granite peaks, some of which are over 8,600 feet. Snow sometimes falls in this mountainous granite region, but the total amount of precipitation is minimal. Some oases are found around both springs and wells.

Because of its rainfall deficiency and correspondingly rugged terrain, Sinai has never boasted a large population. In ancient times the Egyptians were primarily interested in this area for mining turquoise deposits and copper.

History

The recorded history of Egypt began around 3100 BC, when Upper and Lower Egypt were united into one country. As early as the third millennium BC the Egyptians had divided Upper Egypt into twenty-two nomes, or districts, and by the late first millennium BC twenty delta nomes were added to the total.

During periods when the central government was relatively weak, the rulers of the nomes — the nomarchs — were often powerful.

Dynasties	Periods	Approximate Dates (BC)
1st–2nd	Early Dynastic	3050–2700
3rd–6th	Old Kingdom	2691–2176
7th–10th	First Intermediate	2176–2023
11th–14th	Middle Kingdom	2116–1638
15th–17th	Second Intermediate	1638–1540
18th–20th	New Kingdom	1540–1070
21st–25th	Third Intermediate	1070–664
26th	Saite Renaissance	664–525
27th–31st	Late Dynastic	525–330
	Conquest of Alexander	332
	Macedonian Domination	332–304
	Ptolemaic Dynasty	304–30
	Roman Conquest	30

Mark Connally

▲ *Nile Valley near Luxor: note the Late Kingdom temple, lower right, the irrigated fields, the Nile (flowing from right [south] to left, and Luxor and desert on the far east side of the Nile.*

Historians divide the line of kings into thirty or thirty-one "dynasties." Modern historians usually begin the First Dynasty at about 3100 BC and end the series with the Ptolemaic Dynasty (ca. 30 BC). In addition, Egyptologists and historians combine these dynasties into more comprehensive periods or eras: "Kingdom Periods" (Old, Middle, New) were periods of strength and stability while "Intermediate Periods" (First, Second, Third) were periods of disorder and political chaos.

THE GEOGRAPHY OF SYRIA AND LEBANON

The area now occupied by Lebanon and Syria was not only an important region in its own right but also served as a crossroads that connected Babylonia and Assyria with Anatolia (modern Turkey) to the northwest, with the Mediterranean to the west, and with Israel and Egypt to the southwest.

The region is bounded on the west by the Mediterranean Sea, on the north by the Amanus and Malatya Mountains, on the east by a north – south line drawn through Jebel Sinjar, on the south by the Syrian Desert, and eventually on the southwest by Damascus and the Litani River. It is difficult to find a single ancient name that refers to this whole region, although the area west and south of the Euphrates was called "Amurru" (the "West Land"), the "land beyond the River [Euphrates]," and "Aram" during the second and first millennia BC.

Many varieties of landscapes and lifestyles are found in this region. Grain crops can be grown north of an arc that runs from

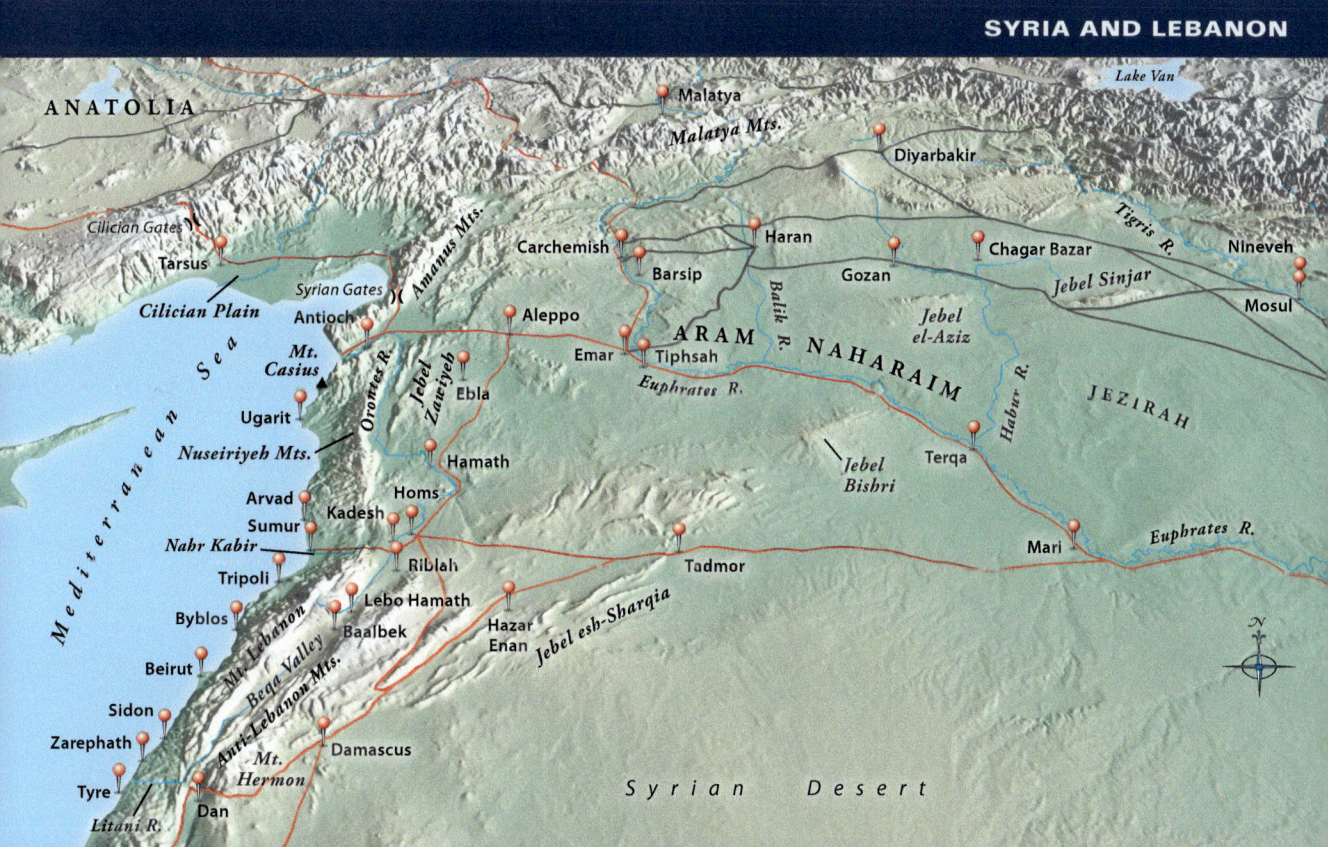

ANATOLIA · Lake Van · Malatya · Malatya Mts. · Diyarbakir · Cilician Gates · Tarsus · Syrian Gates · Carchemish · Haran · Chagar Bazar · Nineveh · Cilician Plain · Antioch · Amanus Mts. · Aleppo · Barsip · Gozan · Jebel Sinjar · Mosul · Mt. Casius · Jebel Zauriyeh · Emar · Tiphsah · ARAM NAHARAIM · Jebel el-Aziz · Ugarit · Ebla · Orontes R. · Euphrates R. · Balik R. · Habur R. · JEZIRAH · Nuseiriyeh Mts. · Hamath · Jebel Bishri · Terqa · Arvad · Homs · Sumur · Kadesh · Mari · Euphrates R. · Nahr Kabir · Riblah · Tadmor · Tripoli · Lebo Hamath · Byblos · Baalbek · Hazar Enan · Jebel esh-Sharqia · Beirut · Mt. Lebanon · Beqa Valley · Anti-Lebanon Mts. · Sidon · Damascus · Zarephath · Mt. Hermon · Tyre · Dan · Litani R. · Mediterranean Sea · Syrian Desert

Mark Connally

▲ *Cedar of Lebanon on snow-covered mountain slopes*
▼ *The island city of Arvad. Tyre was also an island city like this until the days of Alexander the Great (ca. 332 BC), when he linked it to the land.*

Mark Connally

this area. From Carchemish caravans and/or armies could head northwest into Anatolia, westward to the Mediterranean, or southward into Syria, Israel, and Egypt.

Situated in the southwest corner of this whole area is the oasis of Damascus. This city was a key city to Israel, for nearly all traffic entering or exiting Israel from the north had to pass through it. Because of this, the control of Damascus has been much disputed throughout history; yet rarely has Damascus been able to extend its control far in any direction, for it is hemmed in by mountains to the west and north and by desert and basalt outflows to the east and south.

The narrow Mediterranean coastline boasts a series of ports, including Tyre and Arvad (island anchorages), Tripoli, and Ugarit. Immediately inland are the majestic mountains of Lebanon, rising in places to heights of over 10,000 feet and covered with snow six months of the year. This whiteness may have given rise to the name "Lebanon," which is related to a Hebrew root meaning "white." It was here that the prized "cedars of Lebanon" grew.

On the east the mountains of Lebanon drop off into the long and narrow Beqa ("Valley"), where gardens flourish, with olive and fruit trees, vines, and grains. The Orontes River drains the valley to the northeast, and the Litani drains it to the southwest. The Litani River passes through a steep, narrow gorge and flows into the Mediterranean just north of Tyre. East of the Beqa are the Anti-Lebanon Mountains, which were covered with thick forests in antiquity. Because of these obstacles the main international route passed to the east through Damascus.

Close to the north end of the coastal plain, Mount Casius forms a prominent landmark along the shore of the Mediterranean. Just to the north, the plain of Antioch provides a swampy but adequate connecting route from the Mediterranean to Aleppo on the east. The steep scarp of the Amanus Mountains (ca. 6,000 to 7,000 ft.) rises to the north of the plain, and through these mountains a pass leads to the Cilician Plain and on to Anatolia.

Damascus to Jebel Sinjar in the northeast. The land to the west and north of this rough line receives at least 10 inches of rain annually, while rainfall drops off rapidly to the south of this arc. There lie the expanses of the Syrian steppe/desert, where nomads wander with their herds in search of winter grasses. When political conditions were relatively stable, a caravan route across the steppe/desert from Mari to Tadmor became important. These caravans could continue almost due west to Mediterranean seaports near the Kabir River or south to Damascus.

To the north and east of the Euphrates is a steppe area drained by the Habur and Balik Rivers. The roads connecting Assyria, and even Babylonia, with Carchemish ran through

THE GEOGRAPHY OF MESOPOTAMIA

The Tigris and Euphrates Rivers dominate life at the eastern end of the Fertile Crescent. The name Mesopotamia is derived from the Greek and means "land between the rivers." Originally it may have referred to the land between the Euphrates and the Habur Rivers, for the term is used in the Greek translation of the Old Testament (Septuagint) to refer to Aram Naharaim (NIV; lit., "Aram of the two rivers"), which was located near Nahor (Gen 24:10). Both Polybius (second century BC) and Strabo (first century AD) use "Mesopotamia" to refer to the area between the

▼ *The Euphrates at Dura Europos*

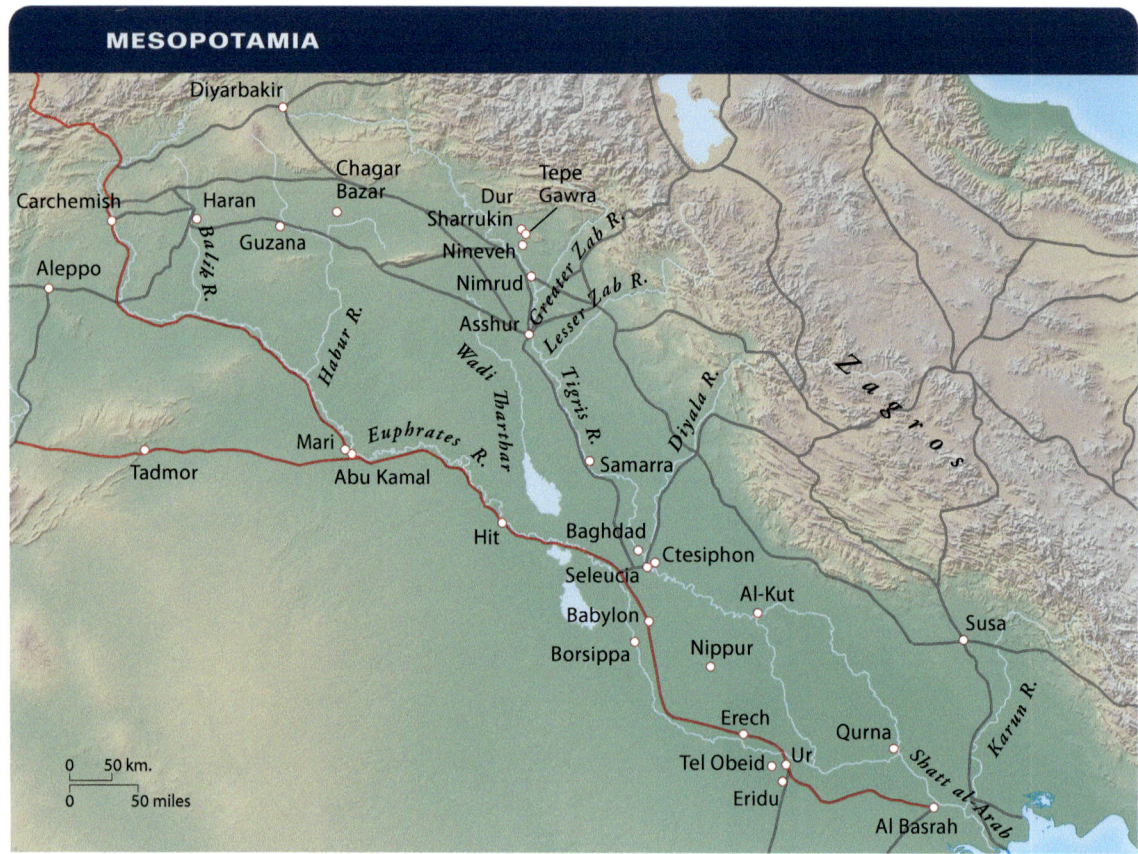

MESOPOTAMIA

Diyarbakir
Carchemish
Haran
Chagar Bazar
Dur Sharrukin
Tepe Gawra
Aleppo
Guzana
Nineveh
Nimrud
Asshur
Balkh R.
Habur R.
Greater Zab R.
Lesser Zab R.
Wadi Tharthar
Tigris R.
Diyala R.
Z a g r o s
Mari
Euphrates R.
Tadmor
Abu Kamal
Samarra
Hit
Baghdad
Seleucia
Ctesiphon
Al-Kut
Babylon
Nippur
Susa
Borsippa
Karun R.
Erech
Qurna
Tel Obeid
Ur
Eridu
Shatt al-Arab
Al Basrah

0   50 km.
0   50 miles

Euphrates and the Tigris. Today it refers to the land between and beside these two great rivers.

The Euphrates and Tigris both have their origins in the mountains of Armenia. Although the source of the Tigris is within a few miles of where the Euphrates passes, the two rivers diverge and follow different paths. The Euphrates is over 1,780 miles long. It begins in Turkey and, after flowing through the mountains of Armenia, it heads due south and joins the North Syrian Plain near Carchemish. It then turns southeast and descends into the Persian Gulf. Since this area receives only 4 to 8 inches of rain each year, agriculture is primarily confined to the narrow river valley, although some grain crops are grown on the steppes, where flocks of sheep and goats feed on the winter grass cover.

The Tigris, which begins in Lake Hazar, is 1,150 miles long. It flows south, through the old heartland of Assyria, and great cities (including Nineveh and Asshur) were once located along its banks. Traditionally grain crops are grown in this area — nourished by the winter rains. Further south near Samarra, there are few canals, scant rainfall, and little habitation.

After Samarra, however, canals branch off the Tigris, and the "delta" region of the Tigris and Euphrates begins. This plain is triangular. The low-lying landscape is flat, expansive, and treeless. The silt, deposited over the millennia, is 15 to 25 feet deep.

This area receives only 4 to 8 inches of rain each year, and agriculture is dependent on irrigation techniques. The channels of both rivers are basically above the surface of the surrounding plain, and if the rivers overflow in the spring, vast stretches of the plain are inundated. The amount of water brought down each year is erratic, so that some years see devastating floods and others disastrous droughts. Thus the residents of the area have tried to harness the rivers by diverting floodwaters to low-lying areas upstream to avoid flooding downstream.

Internally, Mesopotamia produced enough food to feed its population. Land transport was mainly by foot or donkey, but in the southern plain it was impeded by the need to cross the Tigris and Euphrates Rivers, as well as by the numerous canals and channels. Moreover, the plain was often covered by mud.

The main thoroughfares for travelers going from northwest to southeast were the rivers and canals. Bulk goods (such as timber and stone) were transported down the Tigris and Euphrates on rafts supported by animal skins. After the trip, the wooden frames were sold and the skins packed onto donkeys for the return trek northward. Since in antiquity bridges were almost unknown, people used rafts and large circular baskets covered with bitumen to cross the rivers and canals.

Because Mesopotamia lacked many raw materials, it was necessary to import them. Tin was imported from Iran, Afghanistan, and the Caucasus regions; silver from the Taurus Mountains; common timber from the Zagros Mountains; prized cedar wood from the Lebanon and Amanus mountains; and copper from many areas. Also, luxury items were imported from India (spices and cloth) and south Arabia (frankincense and myrrh).

One of the main routes of international as well as local significance that passed through northern Mesopotamia began at Nineveh and ran westward to Carchemish (see p. 23 for the rest of the travel routes). In addition, an important route led northwest along the Euphrates, from the delta toward Mari and then toward Carchemish or Tadmor.

Although remains of human occupation in Mesopotamia date back at least to the Neolithic period (ca. 8000 to 4000 BC),

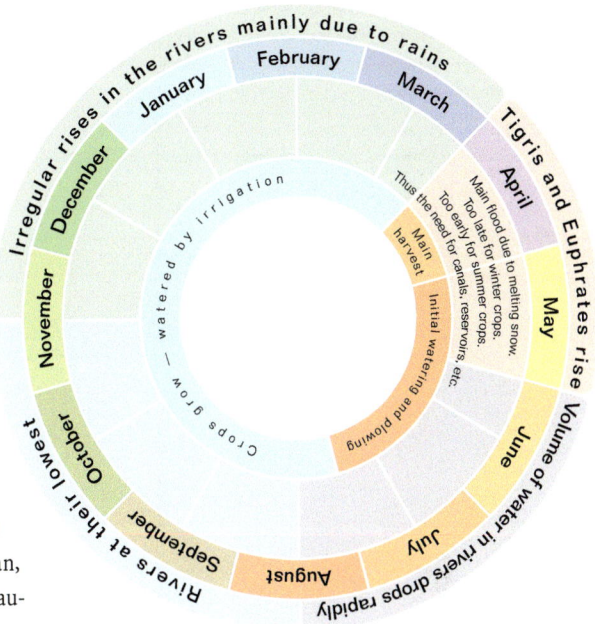

ANNUAL CYCLE IN SOUTHERN MESOPOTAMIA

Mesopotamia, like Egypt, entered the light of history at the beginning of the Early Bronze Age (ca. 3150 BC).

Since life in Mesopotamia can be characterized as a "mud culture" — crops were grown in mud; houses, palaces, temples, ziggurats, etc., were built of mud — it was only natural that mud or clay tablets were used as a medium for communication. By 3100 BC cuneiform (wedge-shaped) writing had developed in Mesopotamia. Because the clay hardens when dried or baked, thousands upon thousands of cuneiform documents have been discovered in Mesopotamia, Armenia, Anatolia, Syria, Israel, and even Egypt.

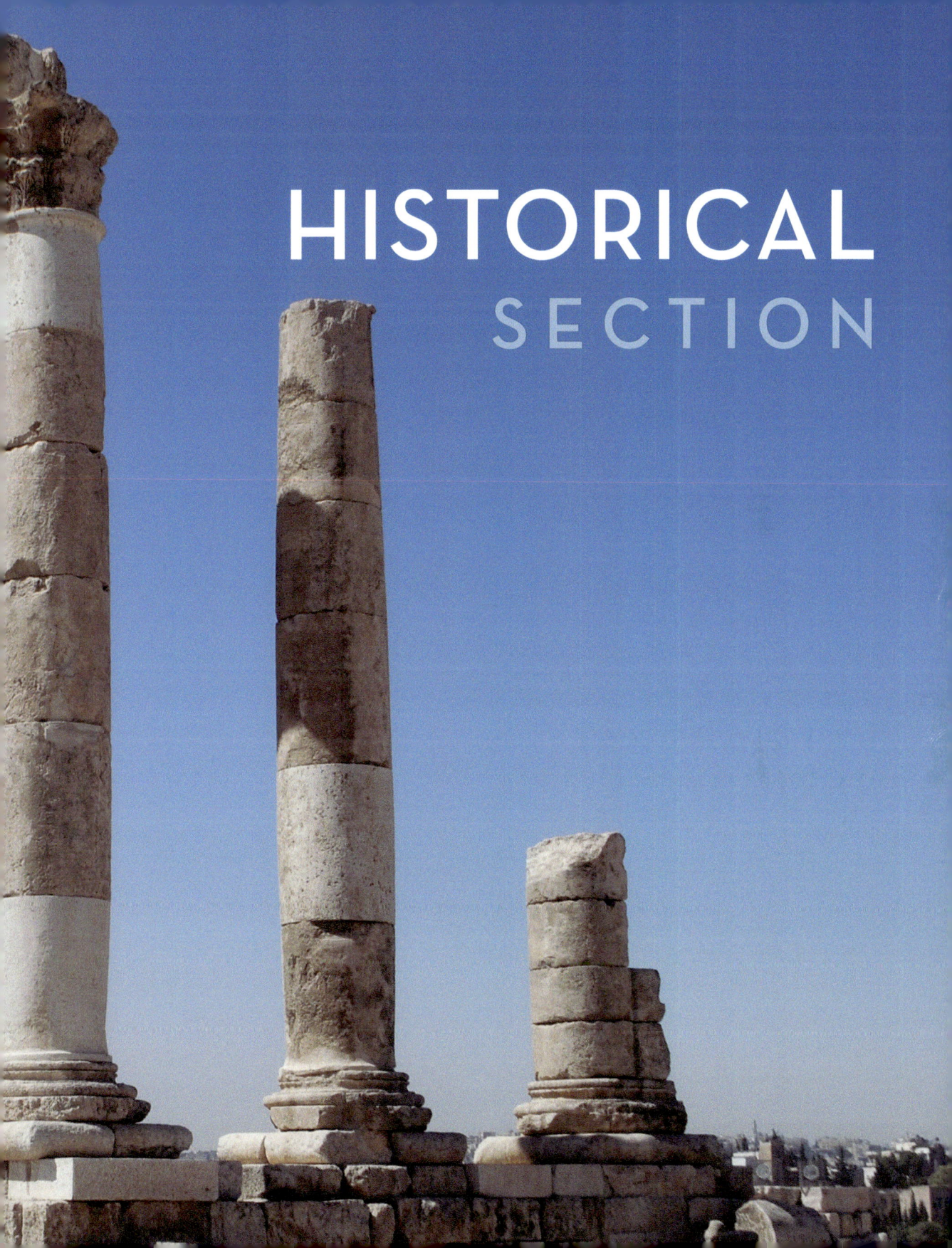

HISTORICAL
SECTION

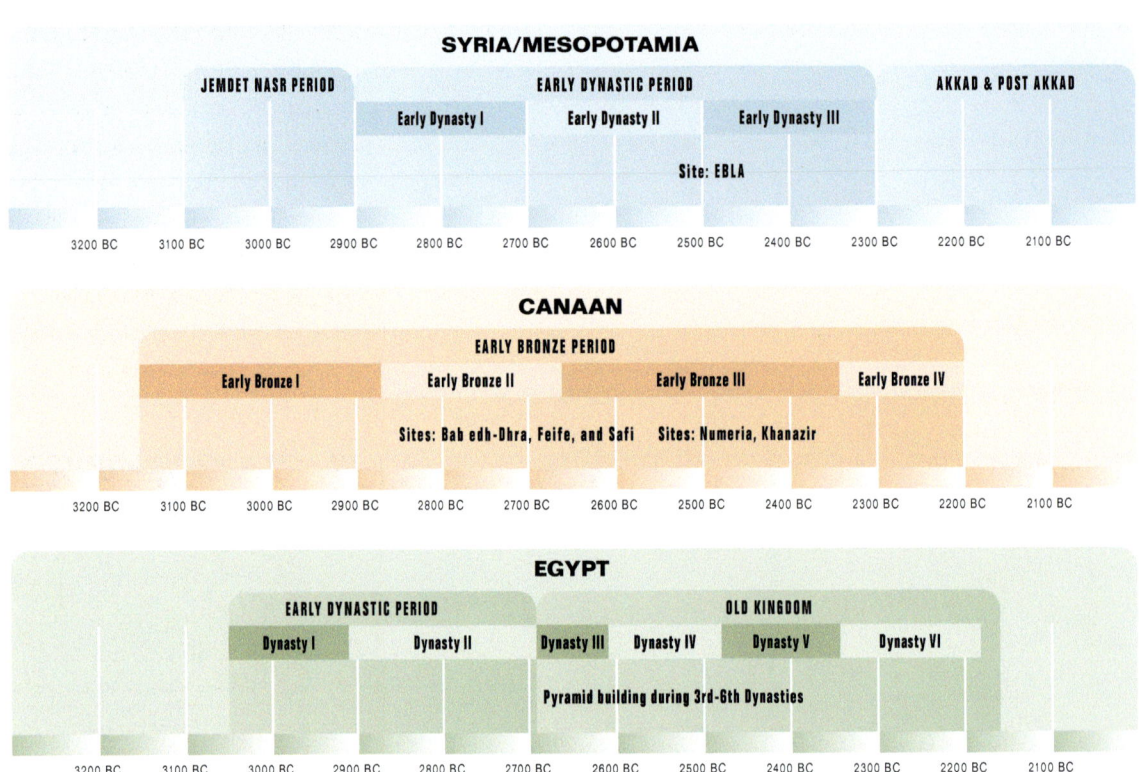

SYRIA/MESOPOTAMIA

JEMDET NASR PERIOD		EARLY DYNASTIC PERIOD			AKKAD & POST AKKAD
	Early Dynasty I	Early Dynasty II	Early Dynasty III		
			Site: EBLA		

3200 BC 3100 BC 3000 BC 2900 BC 2800 BC 2700 BC 2600 BC 2500 BC 2400 BC 2300 BC 2200 BC 2100 BC

CANAAN

EARLY BRONZE PERIOD

Early Bronze I	Early Bronze II	Early Bronze III	Early Bronze IV
	Sites: Bab edh-Dhra, Feife, and Safi	Sites: Numeria, Khanazir	

3200 BC 3100 BC 3000 BC 2900 BC 2800 BC 2700 BC 2600 BC 2500 BC 2400 BC 2300 BC 2200 BC 2100 BC

EGYPT

EARLY DYNASTIC PERIOD		OLD KINGDOM		
Dynasty I	Dynasty II	Dynasty III	Dynasty IV	Dynasty V	Dynasty VI
		Pyramid building during 3rd-6th Dynasties		

3200 BC 3100 BC 3000 BC 2900 BC 2800 BC 2700 BC 2600 BC 2500 BC 2400 BC 2300 BC 2200 BC 2100 BC

THE PRE-PATRIARCHAL PERIOD

Garden of Eden (Genesis 1 – 3)

The book of Genesis, using terse yet picturesque language, recounts the history of the world from the time of creation until Jacob's journey into Egypt. The second section of the creation account (Gen 2:4 – 25) focuses on the creation of the first humans, who were placed in a perfect environment called the garden of Eden.

The name Eden in Hebrew means "delight." Some scholars have suggested that Eden is related to the Sumerian/Akkadian word *edin(u)*, meaning "plain," and that this is a description of the location of Eden. It seems that Eden was the name of a locality, and the special garden planted in the eastern portion of it (Gen 2:8) is described as containing trees and an abundance of water. We can infer that it was in a warm climate — note the mention of fig trees and the lack of clothing (3:7). The major river that watered Eden separated into four headwaters, called Pishon, Gihon, Tigris, and Euphrates (2:10 – 14). The latter two rivers are well known, but the identities of the first two are not.

One view places Eden in eastern Turkey near the headwaters of the Tigris and the Euphrates. If this is correct, then the Pishon and the Gihon might be identified with the Araxes and Murat Rivers. Another view places Eden in southern Iraq. It has the advantage of placing the garden in a warmer climate in the vicinity of the two identifiable rivers, the Tigris and the Euphrates. The Pishon and the Gihon could then have been tributaries of the Tigris and/or Euphrates or canals that branched off these rivers.

The identification of Havilah and Cush (Gen 2:11 – 13) is difficult. "Havilah" may have been a region located in Sinai and/or Arabia (25:18), a locality in which there were supplies of gold, aromatic resins, and onyx. The "land of Cush" usually refers to the land south of Egypt but north of Ethiopia, although it might also refer to a portion of Arabia or even a territory in the mountains east of the Tigris River. Even though the exact location of Eden continues to elude modern interpreters, its theological and spiritual significance has certainly been appreciated by both ancients and moderns alike.

The Table of Nations (Genesis 10)

The biblical text describes the progress of the sinfulness of humankind that was judged climactically in the flood (Gen 6–9). Genesis 10 describes how the nations of the then-known world were derived from Noah's three surviving sons — Japheth, Ham, and Shem. Their seventy descendants mentioned in Genesis 10 are considered to have been the ancestral heads (= eponyms) of the clans and nations that bore their names (v. 32). The list is divided into three major sections, beginning with those peoples farthest removed from Israel's horizon and moving toward her nearest neighbors.

First, the fourteen descendants of Japheth are listed (10:2–5); special attention is given to the "sons of Javan," with whom the Israelites came most often into contact.

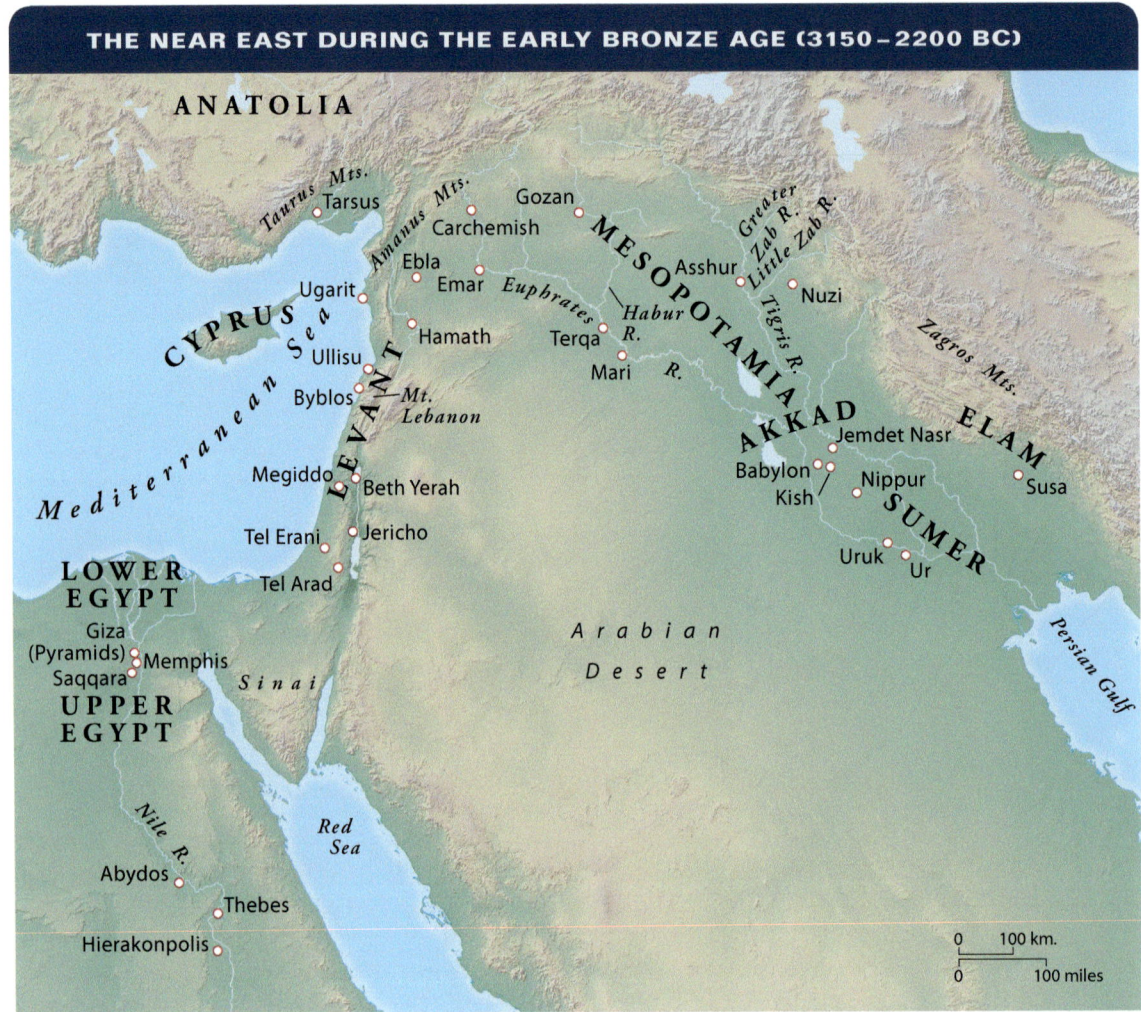

Genesis 10:6–20 lists the thirty-one descendants of Ham. In general, these descendants settled in and around the land of Israel but also to the southeast and in Africa. The list gives special attention to the descendants of Canaan (vv. 15–19), with whom the Israelites came into close contact, and to the cities associated with the warrior Nimrod (vv. 8–12).

The final section (Gen. 10:21–31) contains the twenty-six descendants of Shem. As is typical in Genesis, this list, the most important line, occurs last; Abram/Abraham was a descendant of Shem. The list gives special attention to the descendants of Joktan, people who evidently were the eponymous heads of various Arabian tribes.

Mesopotamia — Early Bronze Age

In Mesopotamia during the Jemdet Nasr Period (3100–2900 BC) writing began, and large cities containing temples, palaces, and fortifications were first built near the Tigris and Euphrates Rivers. Copper tools came into use, and the population grew significantly. This was the era, according to the Sumerian King List, during which the long-lived members of the pre-flood dynasties ruled.

During the Early Dynastic Period (2900–2300 BC) leadership passed between great cities such as Kish, Uruk, and Ur. These city-states had a unified culture, religion, and language, known as Sumerian. The religious center of this civilization was the city of Nippur. During this period much of the Akkadian epic literature originated.

The Early Dynastic Period ended when Sargon of Akkad conquered the old city-states and established the first empire in the region. His grandson, Naram-Sin, seems to have been especially active in the west, even reaching the Mediterranean coast. During this time the arts flourished and literature developed. The Sargonic Empire (2300 – 2100 BC) came to an end because of internal and external pressures.

Egypt during the Early Bronze Age

At the beginning of the Early Dynastic Period (3050 – 2700 BC), Lower and Upper Egypt were united into a single state. Memphis became the capital and remained as such throughout the Old Kingdom Period (2691 – 2176 BC).

The first king of the First Dynasty was Narmer, and his pallet shows him wearing the crowns of both Upper and Lower Egypt. Egypt had contact with Palestine during his reign because pottery inscribed with his name/symbol has

▲ Arad: outline of a portion of the 25-acre Early Bronze Age (2800 BC) town found at Arad. Note the city wall, semicircular towers, streets, and buildings.

▼ Giza, Egypt: pyramids built by rulers of the Fourth Dynasty (ca. 2600 BC), hundreds of years before Abraham

Mark Connally

Map legend:
T. = Tel(l)
Kh. = Khirbet
H. = Horbat

0 10 km.
0 10 miles

Map labels include: T. Dan, T. Anafa, Tel Qadesh, Rosh HaNiqra, T. Kabri, Hazor, Bet Haemeq, H. Yinon, T. Kinneret, T. Keison, Meshed, T. Raqat, Sea of Galilee, Yarmuk R., T. Shimron, Ras Ali, Jokneam, Beth Yerah, T. Sarid, T. Rekhesh, Megiddo, T. Qishyon, Beth Shan, T. Mahfar, Ibleam, Pehel, Jatt, T. Dothan, T. Rehov, T. Hefer, Umm el-Hawa, Jordan R., Kefar Monash, T. el-Far'ah, Jabbok R., Wadi el-Bir, T. Gerisa, T. Aphek, Kh. el-Mahruq, T. Hadid, Gimzo, Et-Tell, Gezer, Jericho, H. Shovav, Jerusalem, Teleilat el-Ghassul, T. Poran, T. Yarmuth, Kh. Ayun Musa, T. Erani, T. Safit, Lachish, Hebron, Kh. Iskander, T. el-Hesi, T. el-Ajjul, T. Agrah, En Gedi, Dibon, Aroer, T. Nagila, Dead Sea, Arnon Gorge, T. Halif, H. Tov, Bab edh-Dhra, En Besor, Arad, Ader, T. Malhata, Numeira, Har Rahama, Zoar, Zered River, Feifa, Khanazir, Mediterranean Sea

thought to symbolize the power of the king as well as to emphasize his association with the sun god Re. They were not only burial places but also religious centers of the cult of the deceased god-king. The pyramids of three of the kings of the Fourth Dynasty — Cheops, Chephren, and Mycerinus, located at Giza — are the most famous.

Although the Old Kingdom is well known because of the pyramids, not many documents have been preserved. We know there were extensive contacts with Nubia to the south, from which luxury items such as gold, ivory, and ebony were obtained. To the north, Egyptian gold work has been found in Turkey, and Egyptian artifacts have been found in Lebanon and Syria.

Egypt began to crumble during the Seventh and Eighth Dynasties. Internal factors, such as the cost of maintaining the pyramid cults, weak kings, and possibly a series of below-normal inundations of the Nile that led to famine seem to have led the country into the turbulent days of the First Intermediate Period.

The Southern Levant during the Early Bronze Age

Although human settlement in the southern Levant has had a long history, it was not until the Early Bronze Age (3150–2200 BC) that we begin to see large urban centers. Strong city walls with protruding defensive towers protected most of the large urban centers. At Arad it is estimated that forty such towers were placed at intervals along the 3,840-foot wall.

Typical houses from this period were rectangular in shape, with the entrance placed close to the center of one of the long walls — hence the name "broad house." The clay model of an Early Bronze house, found at Arad, indicates that the buildings were flat-roofed and windowless. At Megiddo, Arad, and elsewhere, large public buildings, all in the shape of

been found at seven different sites. Archaeology also shows that Egypt had contacts with Mesopotamia during this time.

During the Old Kingdom Period the traditional southern boundary at the first cataract was established, the symbols of hieroglyphic writing were stabilized, administrative structures were put into place, and artistic compositions assumed their stylized forms.

This period is known as the age of the pyramid builders (34 pyramids, out of a total of 47, were built). The pyramids are

▲ *Megiddo: sacrificial altar (25 feet in diameter) from the Early Bronze Age*

▲ *Jericho: neolithic tower 25 feet high discovered by Kathleen Kenyon*

broad houses, appear to have served as temples. At Megiddo four such buildings were discovered in the sacred area along with a large circular altar.

Large burial grounds have been found. At Bab edh-Dhra, for example, it is estimated that some 20,000 tombs contain the remains of 500,000 people. Since there are more burials than the number of people who could have lived in these cities, these cemeteries were probably common burial sites.

In contrast to Egypt and Mesopotamia, no written archives have been found in the southern Levant. One meager reference to the area was found in the tomb inscription of Uni at Abydos in Egypt. He describes how he led five campaigns to the "land of the Sand-Dwellers" during the reign of Pepi I (Sixth Dynasty). The reference to a mountain height situated close to the sea, called the "Antelope-Nose," is probably Mount Carmel.

▼ *Arad: Foundations of a ten-foot-thick city wall and semicircular tower from the Early Bronze Age*

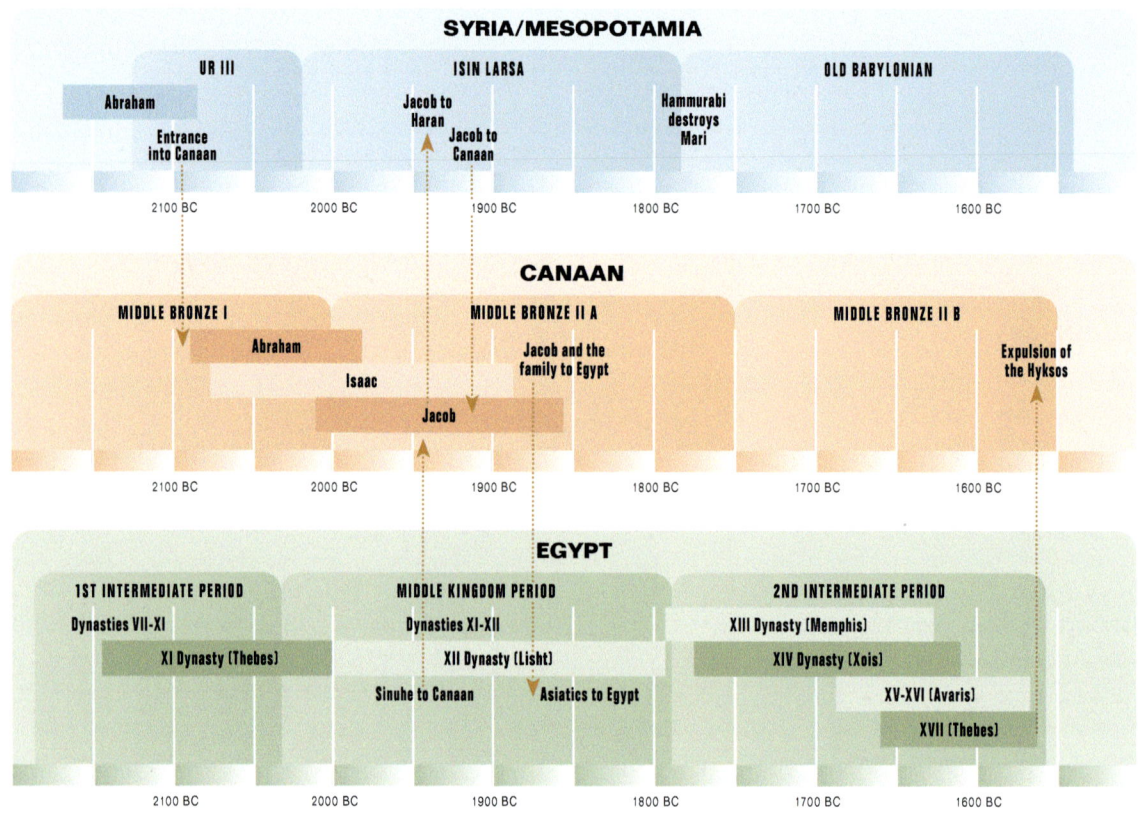

THE PATRIARCHS AND THE EGYPTIAN SOJOURN

The end of the third millennium BC marks the beginning of the era of the biblical patriarchs: Abraham, Isaac, Jacob, and Joseph. Genesis 12–50 records Abram's emigration from Ur of the Chaldeans to Canaan and the events surrounding the lives of the patriarchs. It concludes with the migration of Jacob and his family to Egypt.

The chronological markers in Genesis and Exodus indicate that Abram was born in the late third millennium BC (other chronological systems are also possible). This was a time of relative peace and prosperity in southern Mesopotamia, during which his home city of Ur controlled most of the other city-states in the region. This era, known as the Ur III Period (ca. 2130–2022 BC), is well known from the thousands of cuneiform documents. During this time many ancient Sumerian epics and myths reached their final form. Thousands of economic, legal, and judicial texts witness to the complex and pervasive roles of the palace and the temple in the everyday life of the people. It was from Ur that Abram, at about age seventy, began his earthly pilgrimage of faith (Gen 11:31; Acts 7:4). His first stopping place was Haran, an important caravan city — a trip that took at least thirty days.

Abram stayed in Haran at least a year, for his father, Terah, died there. The roots of Abraham in the Haran/Aram area led to the Israelites referring to their ancestor as a "wandering Aramean" (Deut 26:5).

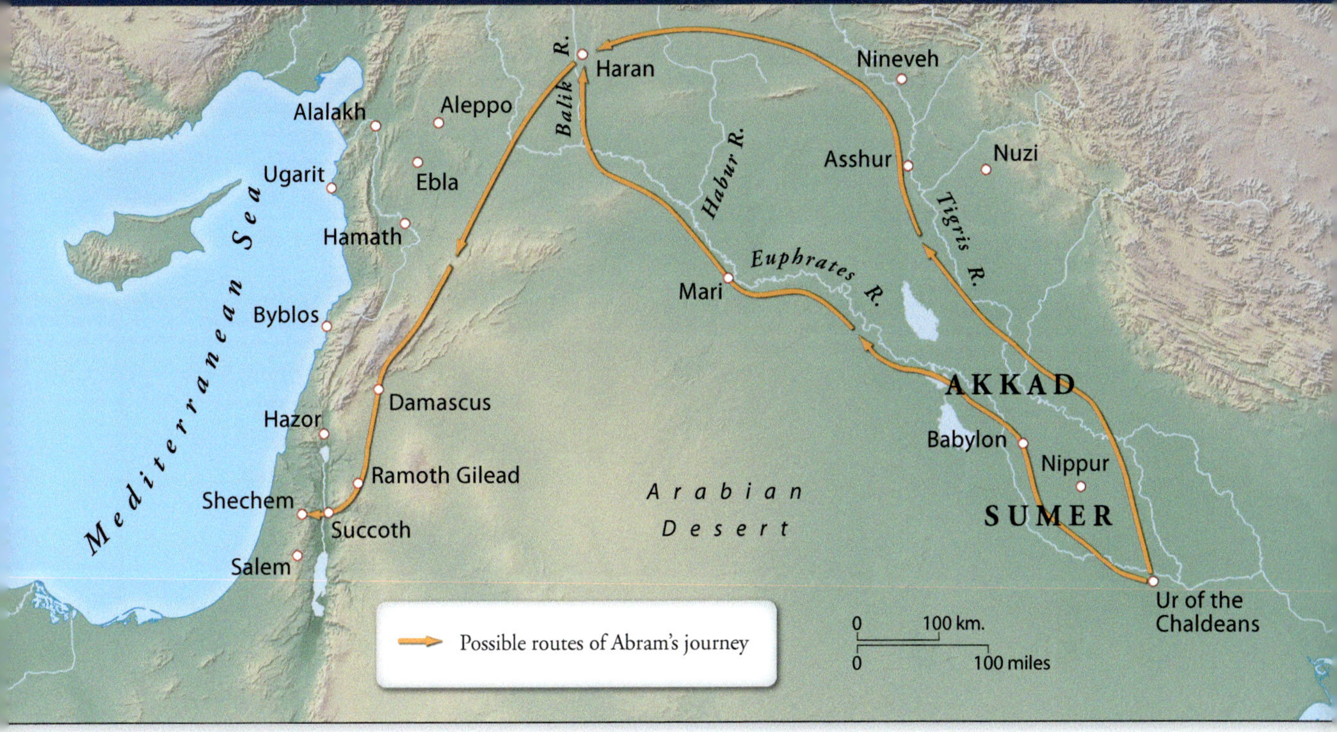

Haran
Balik R.
Nineveh
Alalakh
Aleppo
Asshur
Nuzi
Ugarit
Ebla
Habur R.
Hamath
Euphrates R.
Tigris R.
Byblos
Mari
Mediterranean Sea
AKKAD
Hazor
Damascus
Babylon
Nippur
Shechem
Ramoth Gilead
Arabian Desert
SUMER
Succoth
Salem
Ur of the Chaldeans

→ Possible routes of Abram's journey

0   100 km.
0   100 miles

At age seventy-five Abram "set out for the land of Canaan" (12:4 – 5). His route probably took him south through Damascus to Ramoth Gilead. From there he descended into the Jordan Valley to Succoth, crossed the Jordan River into Canaan, and entered the Hill Country of Manasseh, which took him to Shechem. This trip took at least twenty travel days. This same basic route was later used by Abraham's servant to secure a bride for Isaac (Gen 24), by Jacob as he fled from his brother Esau to his uncle Laban, who lived in Paddan Aram (Gen 27 – 29), and again by Jacob as he returned to Canaan (see esp. 31:19 – 33:20).

At Shechem, at the site of the "great tree of Moreh," the Lord appeared to Abram and promised, "To your offspring I will give this land." In response Abram built an altar and worshiped God there (Gen 12:6 – 7).

From Shechem Abram traveled south to the hills east of Bethel and west of Ai. There he pitched his tent and constructed another altar (Gen 12:8). The topographic details given here fit well by identifying Bethel with modern Beitin and patriarchal Ai with et-Tell. Abram continued southward through the Hill Country of Judah to the Negev (12:9). The

route he traveled from Shechem to Beersheba/Negev is often called the "Ridge Route" or the "Route of the Patriarchs."

If Abram entered Canaan during the Middle Bronze (MB) I Period (2200 – 2000 BC), people were living in tents and huts.

▼ *Et-Tell: looking east at the ruins of the Early Bronze Age site of patriarchal Ai. Possibly Abraham pitched his tent near where this photo was taken (Gen 12:8).*

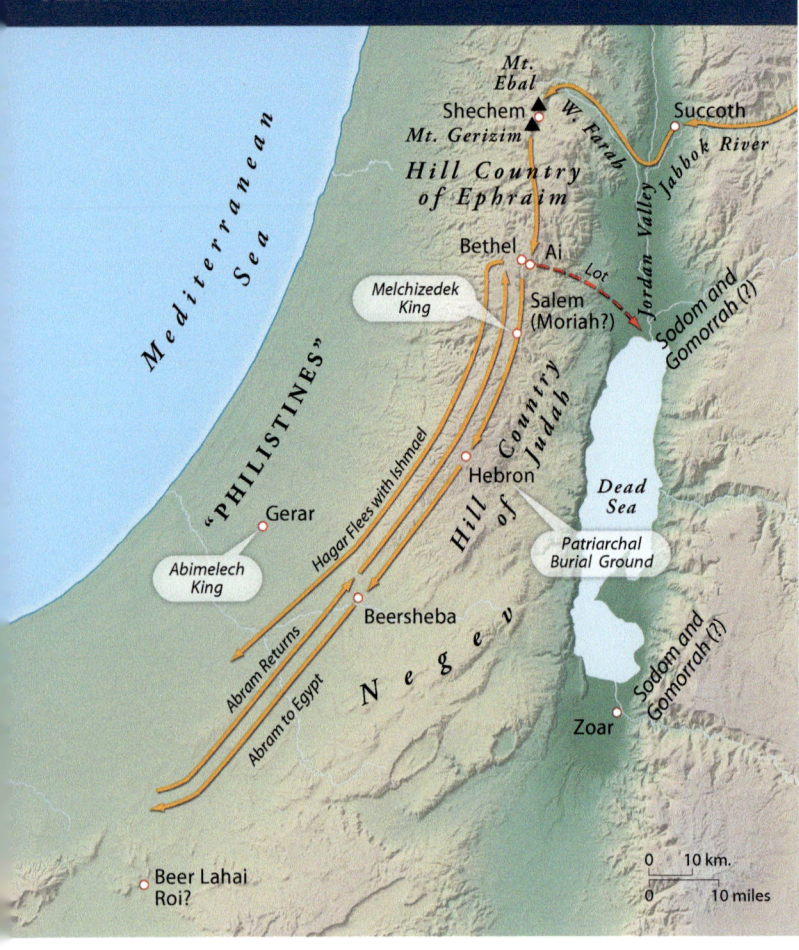

The typical settlement consisted of a cluster of small, flimsy, circular or rectangular installations grouped around a central courtyard in small unwalled settlements. No fortifications or public buildings have been discovered in Palestine for this period. Evidently the Hill Country of Judah was not heavily settled.

Archaeologists have uncovered many MB I tombs. They were usually shaft tombs that led to one or more chambers, with one burial per chamber as a rule. In the Golan region, Transjordan, and elsewhere, fields containing hundreds of dolmens have been found (see photo). These low, tablelike structures were built out of three or four large rocks and sometimes marked shallow gravesites.

Soon after Abram entered Canaan, the land experienced one of its occasional droughts. Abram crossed the northern Sinai Peninsula to Egypt, where he found sustenance for his family (Gen 12:10–20). This was the turbulent First Intermediate Period in Egypt. During this time, the wealthy were consigned to menial tasks while the poor became their masters, the tombs of the kings were plundered, low inundations of the Nile occurred, and death and destruction were everywhere.

▼ *Negev: reconstructed campsite from MB I Period. Probably a circular tent was erected and supported by the partially standing center pole.*

▼ *Golan: a dolmen marking a burial site from the MB I Period (ca. 2200–2000).*

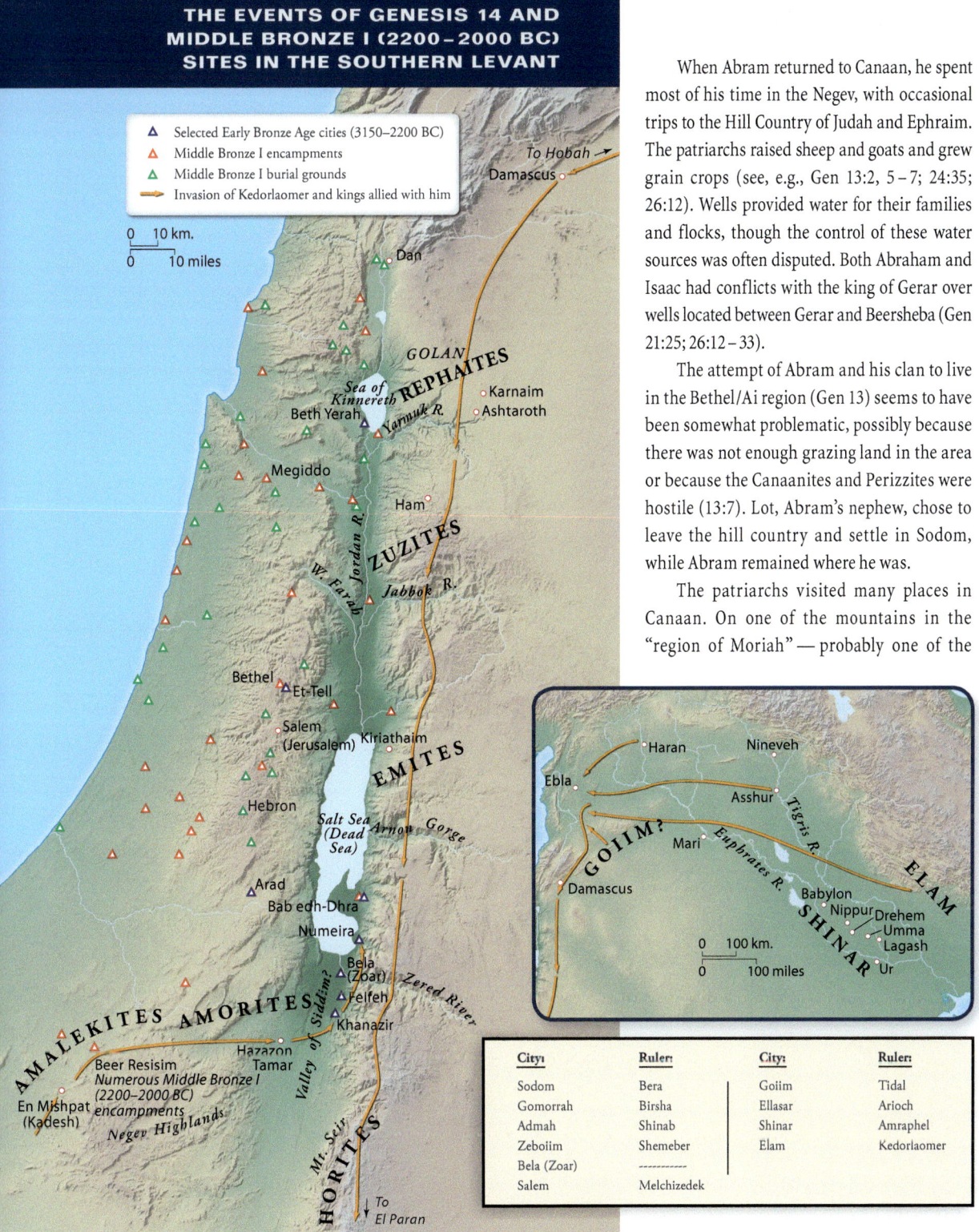

THE EVENTS OF GENESIS 14 AND MIDDLE BRONZE I (2200–2000 BC) SITES IN THE SOUTHERN LEVANT

△ Selected Early Bronze Age cities (3150–2200 BC)
△ Middle Bronze I encampments
△ Middle Bronze I burial grounds
→ Invasion of Kedorlaomer and kings allied with him

0   10 km.
0   10 miles

To Hobah →
Damascus

Dan

GOLAN
REPHAITES
Sea of Kinnereth
Beth Yerah
Yarmuk R.
Karnaim
Ashtaroth

Megiddo

Ham

ZUZITES
W. Farah
Jordan R.
Jabbok R.

Bethel
Et-Tell
Salem (Jerusalem)
Kiriathaim
EMITES

Hebron

Salt Sea (Dead Sea)
Arnon Gorge

Arad
Bab edh-Dhra
Numeira
Bela (Zoar)
Feifeh
Khanazir
Zered River
Valley of Siddim?

AMALEKITES AMORITES
Hazazon Tamar
Beer Resisim
Numerous Middle Bronze I (2200–2000 BC) encampments
En Mishpat (Kadesh)
Negev Highlands
Mt. Seir
HORITES
To El Paran

Inset map

Haran   Nineveh
Ebla
Asshur
Tigris R.
GOIIM?
Mari
Euphrates R.
ELAM
Damascus
Babylon
Nippur Drehem
Umma
Lagash
SHINAR
Ur

0   100 km.
0   100 miles

City:	Ruler:	City:	Ruler:
Sodom	Bera	Goiim	Tidal
Gomorrah	Birsha	Ellasar	Arioch
Admah	Shinab	Shinar	Amraphel
Zeboiim	Shemeber	Elam	Kedorlaomer
Bela (Zoar)	----------		
Salem	Melchizedek		

When Abram returned to Canaan, he spent most of his time in the Negev, with occasional trips to the Hill Country of Judah and Ephraim. The patriarchs raised sheep and goats and grew grain crops (see, e.g., Gen 13:2, 5–7; 24:35; 26:12). Wells provided water for their families and flocks, though the control of these water sources was often disputed. Both Abraham and Isaac had conflicts with the king of Gerar over wells located between Gerar and Beersheba (Gen 21:25; 26:12–33).

The attempt of Abram and his clan to live in the Bethel/Ai region (Gen 13) seems to have been somewhat problematic, possibly because there was not enough grazing land in the area or because the Canaanites and Perizzites were hostile (13:7). Lot, Abram's nephew, chose to leave the hill country and settle in Sodom, while Abram remained where he was.

The patriarchs visited many places in Canaan. On one of the mountains in the "region of Moriah" — probably one of the

Dan

Abdon
Aczib   T.
         er-Ruweisa
Nahariya T. Kabri
Acco   T. Bira
          T. Kison
T. Qashish   T. Shimron
           Jokneam
Dor   Megiddo
T. Mevorach Taanach
T. Zeror   Dothan   Jenin
T. Hefer   Gath   T. Rehov
         T. Poleg
T. Mikhal   Shechem
          Aphek   (Tirzah)
Joppa   T. Gerisa   Shiloh
Yavneh Yam   Bethel
         Gezer   Gibeon
T. Mor   T. Miqne
Ashdod   Jerusalem
Ashkelon   T. es-Safi   Beth Shemesh
                  Beth Zur
T. el-Hesi   Lachish   Hebron
T. el-Ajjul   T. Nagila
T. Ridan   T. Haror   Tell Beit Mirsim
       T. Jemmeh
T. el-Far'ah (S)   T. Masos
                T. Malhata

Tel Qadesh
Hazor
T. Kinnereth
Sea of Galilee
T. Yin'am
T. Rekhesh
Beth Shan
Pehel
T. el-Far'ah
Jabbok R.
Kh. Marjameh (Ein Samiya)
T. es-Sultan (Jericho)
Dead Sea
Arnon Gorge
Zered River

Mediterranean Sea
Yarmuk R.
Jordan River

T. = Tel(l)
Kh. = Khirbet

0   10 km.
0   10 miles

▲ *Dan: well-preserved gate from ca. 1800 BC — the time of the patriarchs*

plain who had revolted against their rule. The defeated kings, along with Lot, were taken captive. Abram pursued them as far as Dan and Hobah, rescuing Lot and the five kings. Upon Abram's return he met with "Melchizedek king of Salem" (v. 18; Salem = Jerusalem [Ps 76:2]).

Genesis 18 – 19, the account of the destruction of Sodom and Gomorrah, records Lot's escape to Zoar. According to the chronology followed here the destruction of Sodom and Gomorrah took place toward the end of MB I. To date, no MB I sites with which any of the five cities in Genesis 14:2 could be identified have been discovered near the south end of the Dead Sea. But recent surveys and excavations along the Transjordanian foothills east and southeast of the Dead Sea have located five sites from the EB period (3150 – 2200 BC): Bab edh-Dhra, Numeira, Zoar, Feifa, and Khanazir (map, p. 39). Some have wondered if these five sites could be the remains of the five cities of Genesis 14:2. This identification is difficult to maintain, however, for the only period during which all these sites were settled was the EB III Period (ca. 2650 – 2350 BC), three hundred years before the date of the events mentioned in the Bible.

The culture of Isaac and Jacob was that of the MB II Period (2000 – 1550 BC). During the first part Isaac and Jacob were active in the land until Jacob went down to Egypt. In Cannan during this time new urban centers were built with monumental city walls, gates, palaces, and temples. Bronze replaced copper as the choice metal for agricultural implements and weapons. Statues, scarabs, and other artifacts of Egyptian origin are found at major sites in the Levant.

mountains in the Jerusalem area — Isaac was "bound" in order to be sacrificed (Gen 22:9; cf. 2 Chron 3:1). The one exception to temporary settlement was a patriarchal settlement close to Hebron. Here Abraham purchased the cave of Machpelah (Gen 23), where ultimately Abraham and Sarah, Isaac and Rebekah, and Jacob and Leah were buried. Evidently Hebron of Abraham's time was located at Jebel er-Rumeidah, where Bronze Age remains have been found.

Although Lot's choice of living in the well-watered Jordan Valley seemed logical, he faced a number of difficulties and twice had to be saved from death by his uncle Abraham. Genesis 14 relates how four kings from the north (led by Kedorlaomer) invaded the area and made war on the five kings of the

In Egypt, the Twelfth Dynasty, in the period known as the Middle Kingdom, was a time of great prosperity. Pyramids were again being built, administrative and bureaucratic structures were in place, and the arts and letters flourished; indeed, this was the "classical" period of Egyptian literature. Commercial contacts with the Levant, especially with Byblos, were common.

Egyptian contact with the Levant is reflected in the Story of Sinuhe (*ANET*, 18 – 23), a tale of an Egyptian who fled from Egypt to the Levant. He first journeyed to Byblos but then

▼ *Gezer: standing stones from ca. 1600 BC — the time of the Israelite sojourn in Egypt — possibly testifying to a covenant/treaty between local Canaanite tribes*

▲ *Nile Delta: the fertile "land of Goshen," where Israel lived in northeast Egypt*

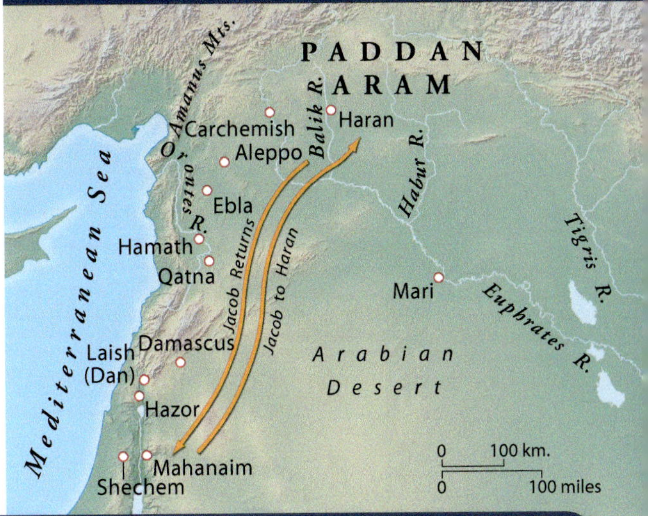

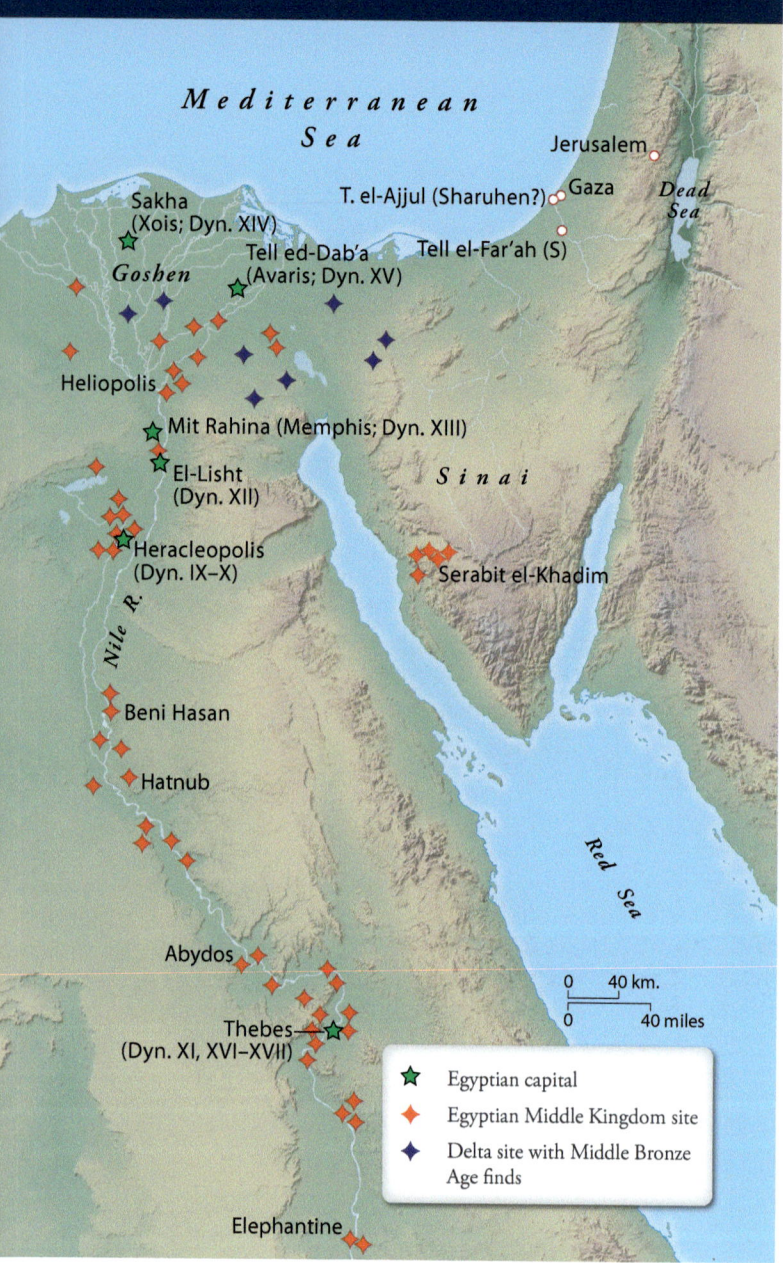

Jacob and his family probably moved to Egypt during the Twelfth Dynasty and settled in the eastern delta of the Nile, in the agriculturally rich land of Goshen, where they remained during their long stay in Egypt (Gen 47:4; cf. Exod 8:22; 9:26).

During the Thirteenth Dynasty, numerous Asiatics from the Levant infiltrated the eastern delta until they became powerful enough to establish what are now known as the Fifteenth and Sixteenth Dynasties. This was the era of the Hyksos domination of the eastern delta. The Egyptians called these Asiatic rulers "rulers of foreign countries," which fits the Hyksos.

During the second part of the MB II Period in Canaan significant urban centers were established (map, p. 40). Temple and palace architecture indicate a high cultural level. The small finds of gold, silver, ivory, and alabaster indicate much prosperity in Canaan. Militarily, important developments include the introduction of the horse-drawn battle chariot and cities protected by glacis (sloping ramps built of packed earth, stone, and plaster) and dry moats.

A good portion of Israel's stay in Egypt took place during the turbulent Second Intermediate Period, and it is possible that the "new king, to whom Joseph meant nothing" (Exod 1:8), was a Hyksos. Exodus 1:9 reads that "the Israelites have become far too numerous for us," which fits much better coming from a Hyksos king — limited in number as the Asiatics were — than a native Egyptian ruler.

At the end of the MB II Period (1550 BC) the Hyksos were expelled from Egypt. Egyptian texts describe battles in the eastern Nile delta, the siege of the Hyksos capital of Avaris, and King Ahmose's driving of the Hyksos rulers out of Egypt and back into Canaan — to the city of Sharuhen (*ANET*, 230 – 34, 553 – 55). This was the area from which kings of the New Kingdom launched their conquests of Canaan.

settled in the land of Araru (possibly in the region of Gilead or Bashan), where he lived until he returned home to Egypt to die. The story even describes in detail the produce of the land — figs, grapes, wine, honey, olives, fruit, barley, emmer (a type of wheat), and cattle — a list that is strikingly similar to that found in Deuteronomy 8:8.

Mark Connally

▲ *Jabbok River: looking at the Jabbok Valley near Penuel (brown mound below the horizon in upper right of the image), near where Jacob met Esau on his return from Haran*

Israel found herself in bondage in Egypt. She had increased in number, yet she did not possess the land promised to Abraham. She had been in Egypt for over three hundred years, but she was anything but a nation. The glowing promises given to Abraham, Isaac, and Jacob may have seemed incapable of fulfillment; yet during the next phase of Near Eastern history, in the face of the most powerful nation on earth, God acted decisively for his people.

THE DELTA

Mediterranean Sea

Sakha
(Xois; Dyn. XIV)

Tell ed-Dab'a
(Avaris; Dyn. XV)

Kom
el-Hisn

Qantir

Farasha

Tell Basta

Tell el-Maskhuta

Tell Atrib

Ghita

Tell es-Sahaba

Nile R.

Tell el-Yahudiya

Heliopolis

Saqqara

Mlt Rahlna
(Memphis; Dyn. XIII)

Dahshur

0   40 km.

0   40 miles

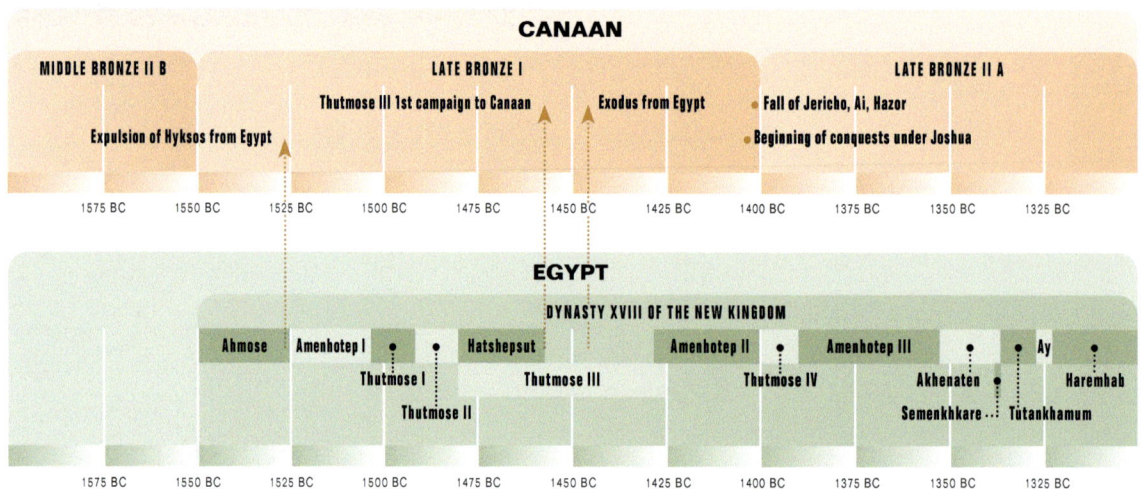

EXODUS AND CONQUEST

▲ *Karnak: Thutmose III slaying his Canaanite enemies held in his left hand. With upraised hands they plead for mercy. Below Thutmose are three rows of cartouches of captive enemies.*

The Exodus from Egypt

One of the most important set of events in the Old Testament centers on the exodus from Egypt, the revelation of God's law at Sinai, and the establishment of Israel in the Promised Land. Since these events (Exod 1 – Josh 11) form a continuous narrative, we will treat them together.

According to the chronology we are using, the exodus and conquest occurred at the end of the LB I Age (1550 – 1400 BC). During the Eighteenth Egyptian Dynasty the oppression of the Israelites came to a climax. The first king of this dynasty, Ahmose, unified Egypt and expelled the Hyksos. Subsequently Thutmose I conducted a military campaign through Canaan into Syria, even reaching the Euphrates River (*ANET*, 234, 239 – 40)! By the end of his reign, Egypt ruled from Syria in the northeast to the fourth cataract of the Nile.

In 1457 BC, at the beginning of Thutmose III's sole reign, he conducted the first of seventeen campaigns into the Levant. His first expedition was his most important one. He defeated a large Canaanite coalition led by the king of Kadesh on the Oron-

▲ Karnak: three cartouches, among many, representing captured enemies of Thutmose III. From right to left the cities of Kadesh (on the Orontes River), Megiddo (in Canaan), and Haszi (in the Beqa Valley) are represented.

▼ Sinai: mountains surrounding Jebel Musa — the traditional site of Mt. Sinai

Gaza Mentioned in Thutmose's first campaign (underlined)

"The capturing of Megiddo is the capturing of a thousand towns."

tes River. By defeating that coalition at Megiddo, Thutmose III took control of most of the southern Levant. No wonder a scribe wrote, "The capturing of Megiddo is [like] the capturing of a thousand towns" (*ANET*, 237). The area Thutmose III ruled included the area where the Israelites would soon begin to settle.

Thutmose III's successor, Amenhotep II, conducted three campaigns into the Levant. The descriptions of these campaigns (*ANET*, 245 – 48) suggest he was losing ground. After his ninth year, Egyptian military activity in the Levant was limited until the days of Seti I.

The oppression of the Israelites begun during the Hyksos era intensified during the early Eighteenth Dynasty (Exod 1:13 – 22). Although not all aspects of Israel's oppression are known, the text indicates that "they built Pithom and Rameses

as store cities for Pharaoh" (v. 11). The city of Rameses was the starting point of the Israelite departure from Egypt (Exod 12:37; Num 33:3, 5) and has been well identified with Tell ed-Dab'a. This huge mound has extensive Hyksos remains and was formerly called Avaris — the capital of the Hyksos.

Although the Bible contains much geographical information in connection with the exodus and the trek to Canaan, the identification of many places remains unknown. This is partly due to the lack of continuity among the populace of the desert-wilderness regions, which makes the preservation of ancient place names almost impossible. Archaeologists have not discovered any artifacts that can be attributed to the nomadic Israelites; living in tents and using animal skins instead of pottery for containers leaves few permanent remains behind.

Thus, scholars are divided in their opinions as to the location even of major landmarks such as the Red Sea and Mount Sinai. There are at least ten different proposals for the location of the Red Sea or Reed Sea and twelve different candidates for Mount Sinai. In spite of these uncertainties, a few suggestions can be made regarding the exodus and wanderings. After leaving Rameses, the Israelites journeyed to Succoth. Note that "God did not lead them by the way of the land of the Philistines" (Exod 13:17 NASB) — the well-known route across northern Sinai to Gaza that was well guarded by Egyptian troops. Since the Israelites were led "around by the desert road toward the Red Sea" (v. 18), they apparently were heading southeast toward modern Suez (13:20 – 14:9).

Next the Israelites crossed the Red Sea. Since the Hebrew text literally means the "Reed Sea," many scholars look for a location in the lake/marsh areas that used to exist in the region through which the Suez Canal now passes. A location near the junction of the Great and Little Bitter Lakes is as plausible as any. According to nineteenth-century travelers, the water at that spot was not very deep, and they even mention that at times, the depth of the water decreased when the wind shifted. Note that according to the text, "the LORD drove the sea back with a strong east wind" (Exod 14:21).

The identification of Mount Sinai (Horeb) with Jebel Musa ("Mount Moses") is based on Christian tradition dating back to the fourth century AD. There, during the Byzantine period (AD 324 – 640), the desert monastery of St. Catherine was established. But the suggested identification of Mount Sinai with Jebel Sin Bisher also deserves careful attention. Some of the biblical data supports this identification. For example, it

SINAI DURING THE BRONZE AGE

is approximately three days' journey away from Egypt (Exod 3:18; 5:3; 8:27) at a desert junction where there were supplies of water. It is close to Egypt near the road that led from Midian to Egypt, and thus it would make a plausible location for the burning bush incident. Moses could have brought Jethro's sheep along this road in order to use the water canal and pasturage found on the eastern edge of the Nile delta when the Lord appeared to him in the burning bush. This took place near the mountain where he would later worship him (3:1).

After camping at Mount Sinai for about a year, the Israelites set out for the Kadesh Barnea region. This trek normally took eleven days (Deut 1:2), which fits the identification of Mount Sinai with Jebel Sin Bisher better than with Jebel Musa. On their way they passed from the Desert of Sinai through the Desert of Paran to the Desert of Zin near Kadesh (Num 10:12; 33:36).

From Kadesh twelve men were sent to "explore the land of Canaan" (Num 13:2). They went up through the Negev into the hill country, traveling as far north as Rehob and Lebo Hamath (13:21). The "land of Canaan" is presented here as a geopolitical entity with definable boundaries that are described in Numbers 34:1 – 12 and Ezekiel 47:13 – 20.

Since the account of the exploration of Canaan gives prominence to the Hebron region and the nearby Valley of Eshcol (Num 13:22 – 24), these Israelite scouts most likely traveled the Ridge Route along the central mountain ridge. Although other cities are not mentioned by name, they are described as being "fortified and very large" (v. 28). The land itself is described as fruitful; note the large cluster of grapes, the pomegranates, and the figs that the scouts brought back to camp (vv. 23 – 24). It was a land "flowing with milk and

THE LAND OF CANAAN EXPLORED BY THE TWELVE SCOUTS

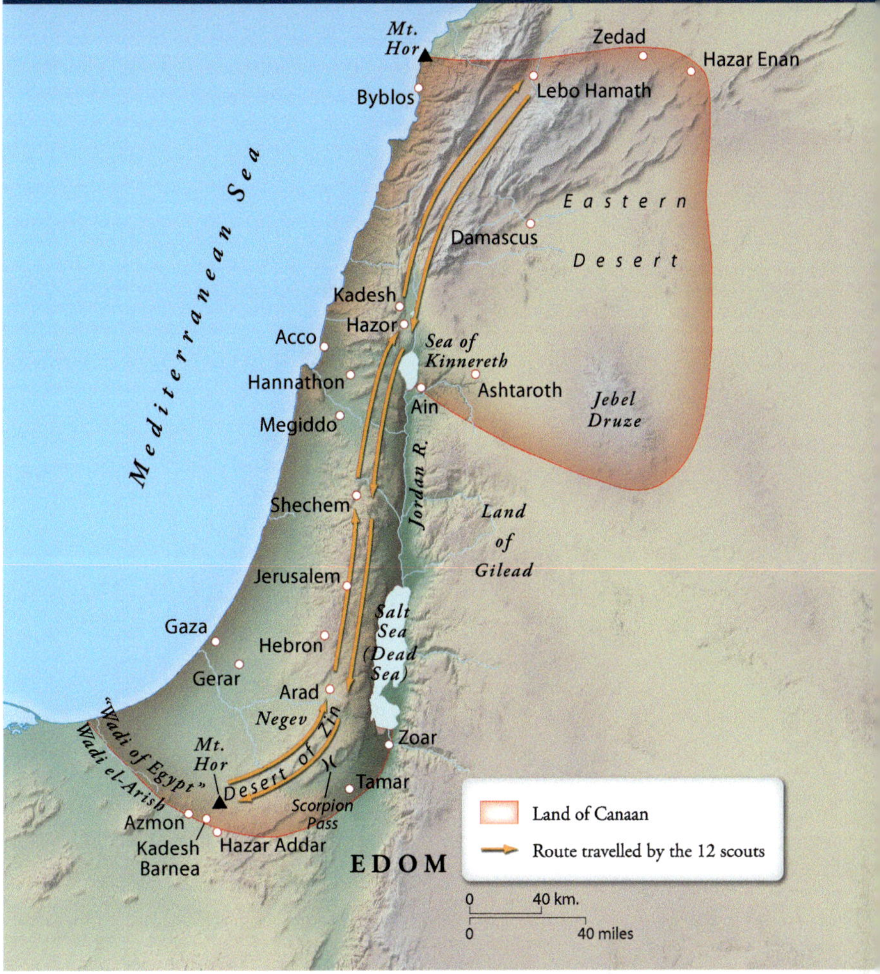

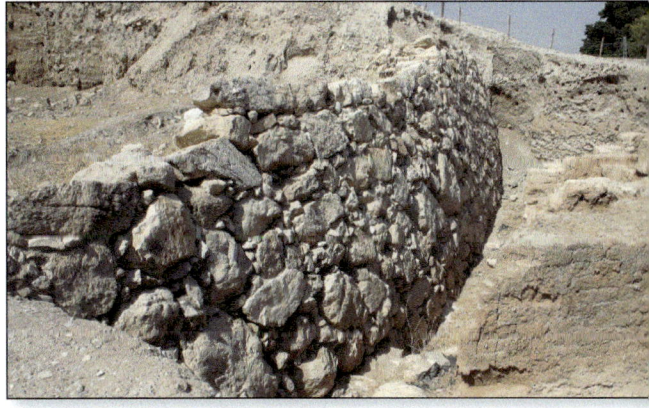

▲ *Jericho: Middle Bronze "revetment wall" upon the top of which one of the city walls stood*

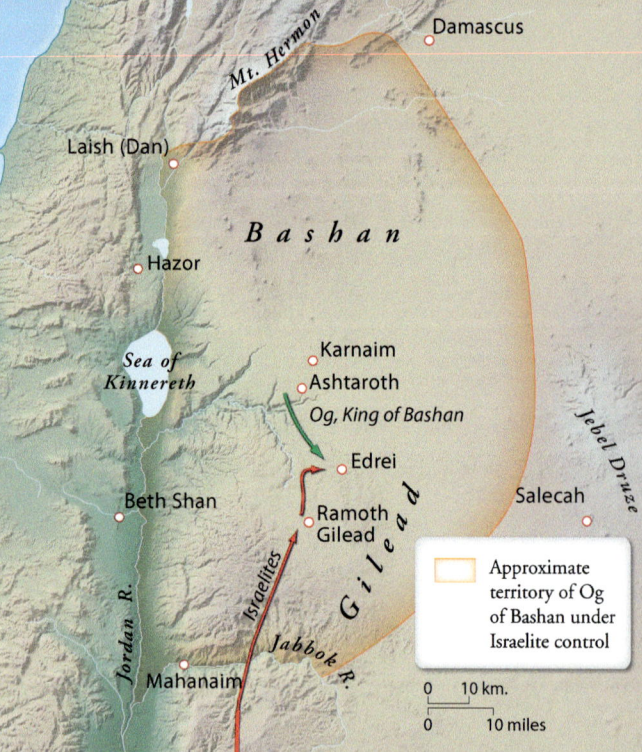

honey" (v. 27) — symbols of bounty and abundance.

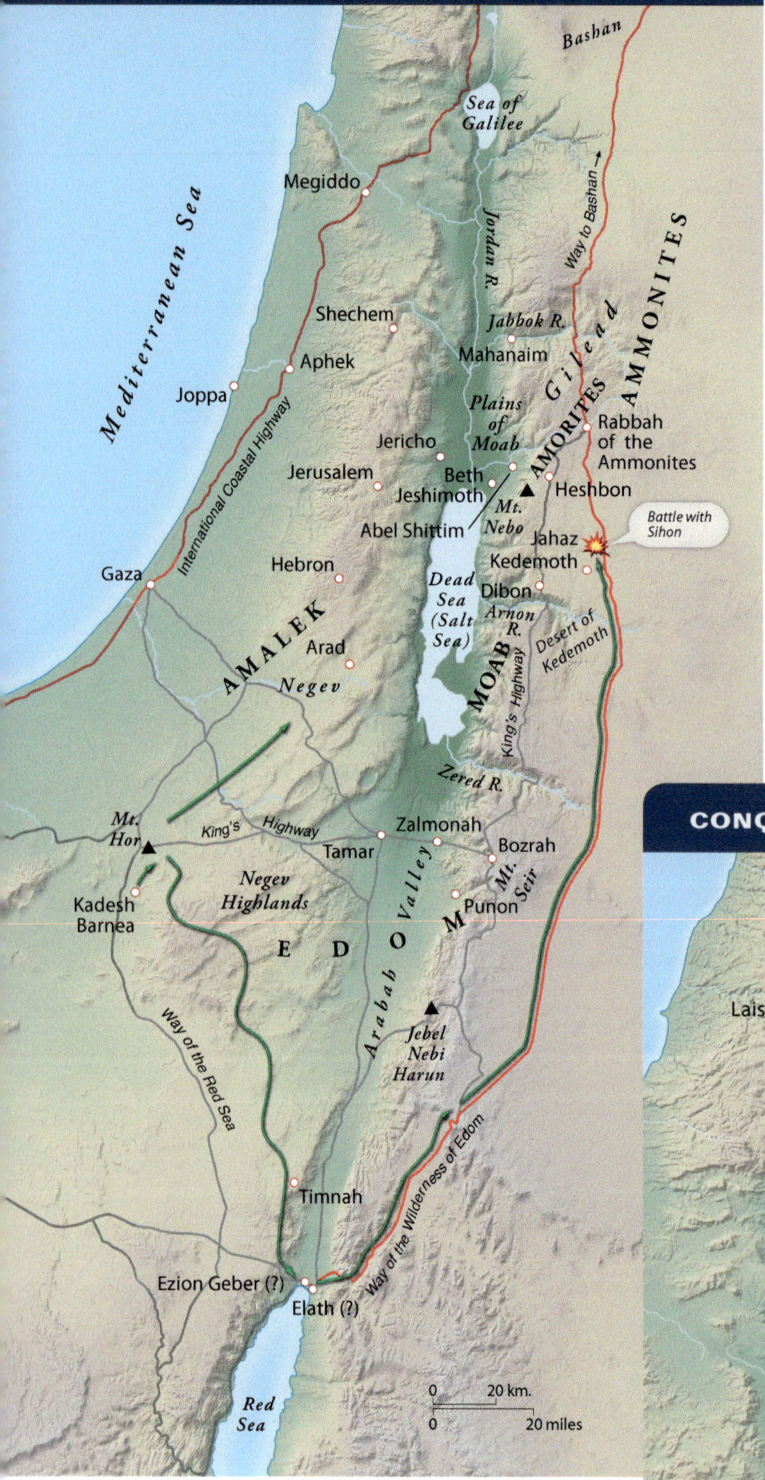

Because of the people's disobedience, God consigned them to wander in the wilderness for forty years. After making an abortive attempt to enter Canaan (Num 14:39–45), the Israelites began their bleak wilderness experience. They spent much of their time in the desolate region between Kadesh and Ezion Geber (near/on the Red Sea [33:36]), likely camping near Kadesh Barnea in the western Negev Highlands.

At the end of this period Miriam, Moses' sister, died (Num 20:1). Near Kadesh Moses disobeyed God's command by striking the rock twice when he should have spoken to it (vv. 2–13); also at Kadesh Moses requested the king of Edom to allow the Israelites to pass through his land along the "King's Highway" (v. 17). Clearly the Edomites had extended their control from the Transjordanian Mountains westward, across the Arabah Valley into the Negev Highlands. In spite of Israel's promise to stay on the highway and to purchase

CONQUEST OF GILEAD AND BASHAN

Approximate territory of Og of Bashan under Israelite control

▲ *Jericho oasis from the mound of Jericho*

water (vv. 17 – 19), their request was refused. Setting out from Kadesh, the Israelites arrived at Mount Hor, where Aaron died and Eleazar was appointed in his place (vv. 22 – 29).

Next the Israelites attempted to enter Canaan (Num 21:1 – 3). At first the king of Arad, who lived in the Negev, defeated them; but after praying to God, the Israelites were victorious in their second encounter. Nevertheless, they followed a circuitous path to the south and east. Their major route seems to have taken them south to the tip of the Red Sea (to Ezion Geber and Elath, see Deut 2:8), then heading north to east of the lands of Edom and Moab (Judg 11:18), traveling on what was called the "King's Highway" (Num 21:22; here the eastern branch of the Transjordanian Highway).

From the "Desert of Kedemoth" (Deut 2:26 – 30) messengers were sent to Sihon, the king of the Amorites (who lived in Heshbon), requesting permission to pass, from east to west, through his territory to the Jordan River. Sihon refused and marched against the Israelites, but he was defeated (Num 21:23 – 24). Thus Israelites took possession of Sihon's land from the Arnon River in the south to the Jabbok River in the north (21:24 – 25; Deut 2:36; Judg 11:22).

Next they marched north "along the road toward Bashan" (Num 21:33). Og, the king of Bashan, who lived in

Ashtaroth (Deut 1:4), battled Israel at Edrei. He was defeated and his territory from the Jabbok River to Mount Hermon came under Israelite control (Num 21:33 – 35; Deut 3:4 – 11).

Numbers 21 through Deuteronomy 34 describes the events that occurred during Israel's encampment, east of Jericho, in the Plains of Moab between Beth Jeshimoth and Abel Shittim (Num 22:1; 33:49). There Balaam blessed rather than cursed the Israelites (Num 22 – 25), and there Moses preached his last sermons before ascending Mount Nebo to die (Deut 34).

The Conquest of Canaan

After Moses' death, under Joshua's leadership, Israel crossed the Jordan east of Jericho. This event took place in the spring of the year, for the Jordan was overflowing its banks after the winter rains and while the (barley?) harvest was taking place (Josh 3:15); also, the Passover (March – April) was celebrated soon afterward at "Gilgal on the plains of Jericho" (5:10).

The first city to be captured by the Israelites was Jericho (Josh 6). This city has been well identified with Tell es-Sultan, a ten-acre mound situated beside a powerful spring in an otherwise arid region. At most 2,000 people lived there. Some

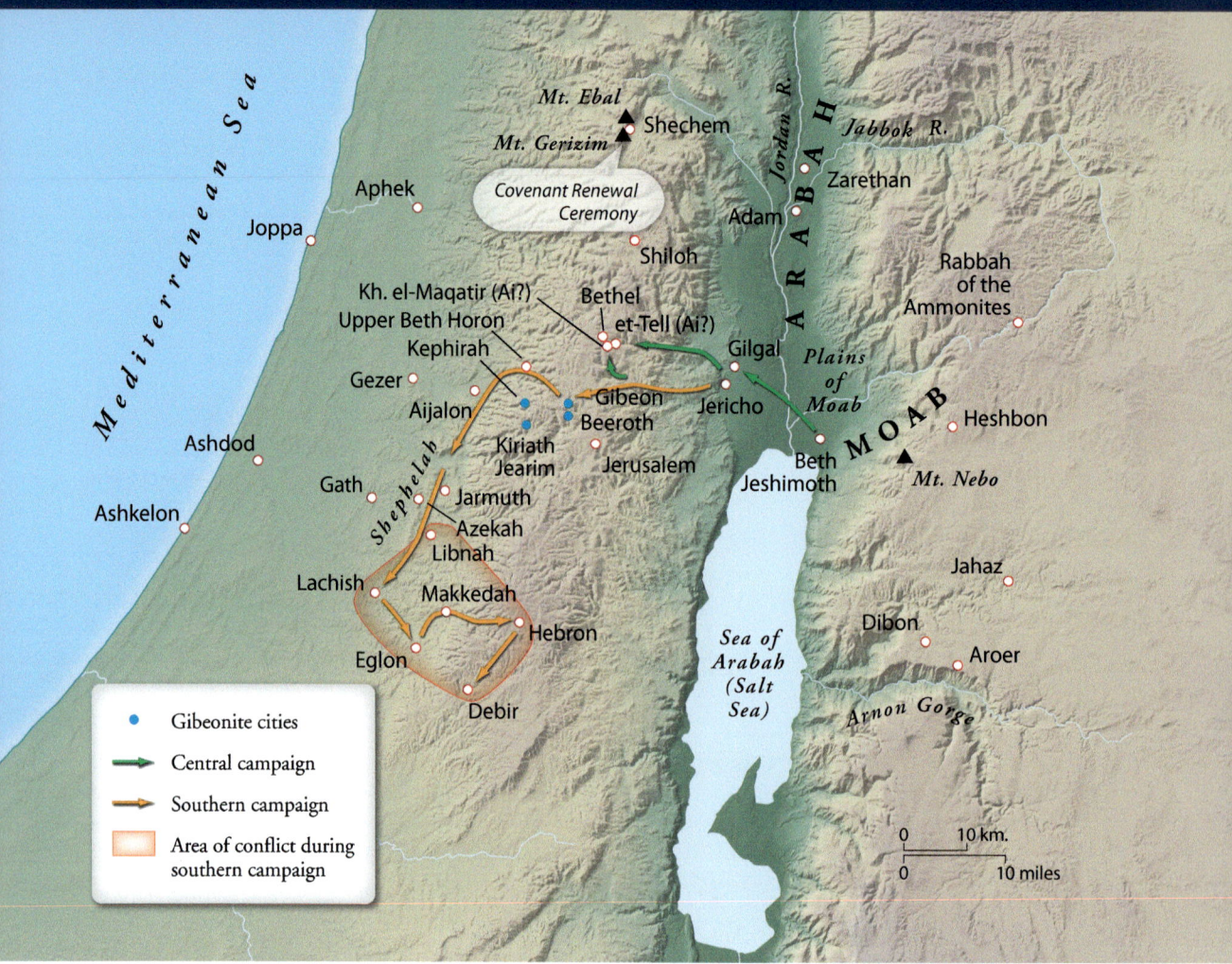

scholars do not believe that the archaeological profile of the site corresponds to the history of Jericho as found in the biblical sources, and indeed, if one holds to a late date for the conquest (ca. 1250 – 1230 BC), no one was living at Jericho at that time. However, if one holds to an early date for the conquest (ca. 1406 BC), the archaeological profile fits much better. Pottery and scarabs from nearby tombs indicate that people were living there at that time.

The conquest of Ai (Josh 7 and 8) presents the historical geographer with serious problems. Biblical Ai is normally identified with et-Tell. This town is said to lie to the west of Bethel (Gen 12:8; Josh 8:9, 12). Thus the usual identification of Bethel with modern Beitin lends support to the identifica-

tion of Ai with et-Tell. But the archaeological picture of et-Tell does not agree at all with the historical data found in the Bible. According to the excavators, the large EB city was destroyed ca. 2400 BC, and the site was not inhabited again until ca. 1200 BC. Thus, no matter which date of the conquest is espoused (whether 1406 or 1250 BC), et-Tell was apparently unoccupied at the time Ai was supposed to have been conquered by Joshua.

In fact, there seem to be three different places called "Ai" in the Bible. The Ai mentioned in the patriarchal narratives (ca. 2000 BC; Gen 12:8; 13:3) is to be identified with et-Tell. The Ai destroyed by Joshua (ca. 1400 BC, Josh 7 – 8) is possibly to be located at the two-acre fortified site called Khirbet el-Maqatir, located only 0.6 mi. to the west of et-Tell and

1 mi. south-southeast of Beitin (= Bethel). A LB I walled fort has been found there. The Ai of Ezra 2:28 and Nehemiah 7:32 (sixth and fifth centuries BC) is to be located at Khirbet Haiyan (about 1 mi. southeast of et-Tell).

Next the Israelites moved northward to the Shechem area. Nearby, on Mount Ebal, Joshua built an altar to God, sacrificed, and wrote on stones (monumental, standing stones?) a copy of the Mosaic law (Josh 8:30 – 35). On Mount Ebal and Mount Gerizim the Israelites read the curses and blessings of the law (Josh 8:33 – 34; see Deut 27:11 – 14) as the covenant between the people and Yahweh was renewed.

When the people of the land of Canaan heard of the Israelite victories at Jericho and Ai, they banded together for mutual defense (Josh 9:1 – 2). However, a group of Hivites living in the cities of Gibeon, Kephirah, Beeroth, and Kiriath Jearim (vv. 7, 17) chose to make a treaty with Israel. At the time they made the treaty, the Israelites thought that the Hivites were "from a distant country" (v. 6). But Israel soon discovered that the Gibeonites lived in the heart of the land.

When the king of Jerusalem heard of the Gibeonite-Israelite treaty, he was alarmed. A major neighboring city (10:2), Gibeon, had gone over to the side of the invading Israelites, and it, as well as its three allied cities, sat astride the two major roads that led from Jerusalem to the coast. Thus Jerusalem's lifelines to the coast and to her Egyptian ally were cut off, and she could become easy prey to the Israelites.

Jerusalem assembled a coalition consisting of the kings of Hebron, Jarmuth, Lachish, and Eglon (Josh 10:5). As these rulers moved against Gibeon, the Gibeonites appealed to Joshua for assistance on the basis of their treaty. Joshua responded by marching all night from Gilgal up into the hill country to relieve the siege of Gibeon. The coalition was defeated, and

▲ *Gibeon: the chief of the four Gibeonite cities that made a treaty with Israel (Josh 9 – 10)*

the kings and their armies fled westward, down the descent of Beth Horon, heading for the safety of their cities located in the Shephelah (Josh 10). The Israelites, with divine assistance — hailstones and a prolonged day — defeated these armies, and the kings themselves were executed at Makkedah (vv. 21 – 27). What began as a rescue mission ended with the conquest of southern Canaan — the second major campaign led by Joshua.

The final phase of the conquest of Canaan — the northern campaign — occurred when Jabin, the king of Hazor, put together a coalition that included the kings of Madon, Shimron, and Acshaph (Josh 11:1 – 3). These kings encamped at the "Waters of Merom" (vv. 4 – 5). The Israelite attack was successful, and as the defeated kings retreated, the Israelites pursued them to the Sidon region. Jabin's city, Hazor, was burned (vv. 10, 13), but the Israelites did not follow up their victory by establishing a settlement at Hazor, for archaeology shows that Canaanites reoccupied the city and lived there until its conquest by Deborah and Barak (Judg 4 – 5), around 1200 BC.

Thus the initial stages of the conquest of Canaan were completed. Yet the biblical writers were well aware that within Canaan there were still large portions of country controlled by non-Israelites (e.g., Josh 13:1 – 7; map p. 59). The apportionment of the land, the settlement of the Israelites in it, and the attempt to deal with the non-Israelite population groups would preoccupy the Israelites for the next several hundred years.

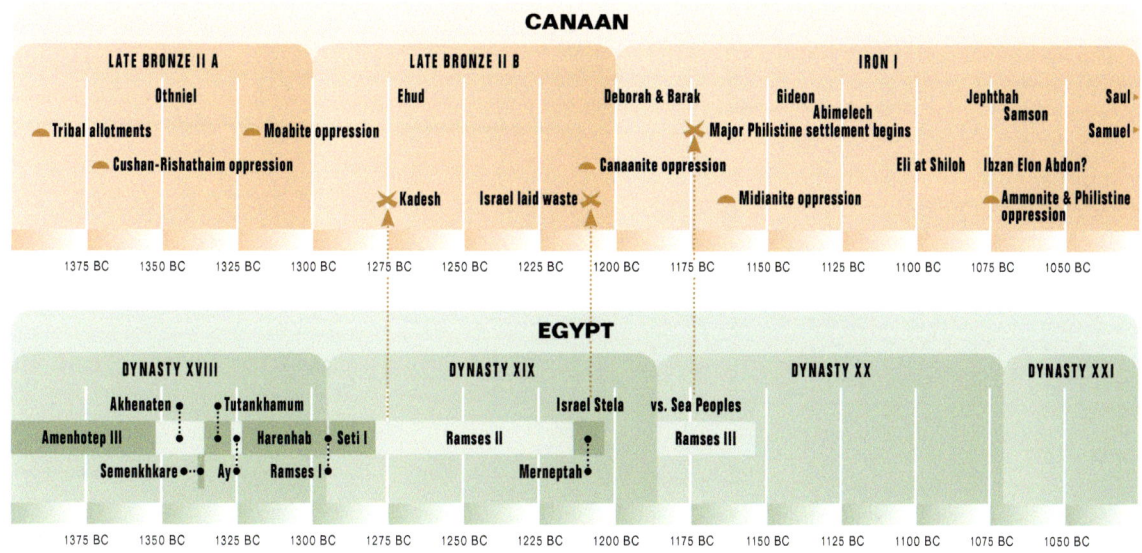

CANAAN

LATE BRONZE II A			LATE BRONZE II B				IRON I				

Othniel — Tribal allotments — Moabite oppression — Cushan-Rishathaim oppression

Ehud — Kadesh — Israel laid waste

Deborah & Barak — Canaanite oppression — Major Philistine settlement begins — Midianite oppression

Gideon — Abimelech — Eli at Shiloh — Jephthah — Samson — Ibzan Elon Abdon? — Saul — Samuel — Ammonite & Philistine oppression

1375 BC 1350 BC 1325 BC 1300 BC 1275 BC 1250 BC 1225 BC 1200 BC 1175 BC 1150 BC 1125 BC 1100 BC 1075 BC 1050 BC

EGYPT

DYNASTY XVIII	DYNASTY XIX	DYNASTY XX	DYNASTY XXI

Akhenaten • Tutankhamum — Amenhotep III — Harenhab • Seti I — Semenkhkare • Ay • Ramses I — Ramses II — Merneptah — Israel Stela — vs. Sea Peoples — Ramses III

1375 BC 1350 BC 1325 BC 1300 BC 1275 BC 1250 BC 1225 BC 1200 BC 1175 BC 1150 BC 1125 BC 1100 BC 1075 BC 1050 BC

SETTLEMENT IN CANAAN AND THE TIME OF THE JUDGES

Allotment of the Land

After the initial conquest of Canaan, the various Israel-ite tribes began to receive territory (see Josh. 13 – 21). Although the actual apportionment took place dur-ing the days of Joshua and Eleazar the son of Aaron (14:1; 19:51), later copyists and editors of the book of Joshua seem to have "updated" the lists of cities mentioned as belonging to the various tribes.

The tribal territories are described in three ways. In some cases the boundary of a given tribe is described as a line run-ning from point A to B to C, etc., in a "dot-to-dot-to-dot" fash-ion (e.g., Judah's boundary in Josh. 15:1 – 12). Second, the text sometimes lists cities belonging to a given tribe (e.g., Judah in

15:21 – 63; Benjamin in 18:21 – 28). A third method notes the cities on the extremities of a tribal territory (e.g., Reuben in 13:16 – 17; "from Aroer … to Heshbon").

▼ *Hill Country of Judah where Israel settled. Terrace farming, rich soil, stone buildings.*

heartland of Judah was confined to the heights of the mountains, but during periods of strength, they expanded westward into the Shephelah and southward into the Negev. Only rarely did Judah control the Philistine Plain.

Simeon (Josh 19:1 – 9; 1 Chron 4:24 – 43)

Simeon received territory inside Judah. Of the seventeen cities mentioned (Josh 19:2 – 7), fifteen were previously mentioned in the city list of Judah. Its primary location was in the western Negev. Since Simeon's territory received only 10 inches of rain each year, the tribe specialized in keeping flocks.

Ephraim (Josh 16; Judg 1:29)

Ephraim's southern boundary matched the northern one of Benjamin (Josh 16:1 – 5; 18:12 – 13), while on the west it was blocked from the sea by the tribe of Dan (19:40 – 48). Its northwestern boundary ran along the Kanah Ravine (16:8; 17:7 – 10), while on the north and east fewer border points are provided. Like Judah, Ephraim lived in the rugged mountainous area. Deep V-shaped valleys provided it with security and inaccessible areas.

Judah (Josh 15; Judg 1:8 – 18)

Judah is the first tribe to be allocated territory. Its southern border (15:1 – 4) was identical with that of the land of Canaan (Num 34:3 – 5; map p. 47), while its northern border coincided first with the southern boundary of Benjamin (Josh 15:5 – 10; 18:14 – 19) and then, following the path of the Sorek Valley westward to the Mediterranean Sea (15:10 – 11), with the southern boundary of the tribe of Dan (map p. 57).

In addition a long list of 132 cities is given for Judah, divided into four major geographical areas: the Negev (Josh 15:20 – 32), the Shephelah (vv. 33 – 47), the hill country (vv. 48 – 60), and the eastern wilderness (vv. 61 – 62). The

Manasseh (Josh 17; Judg 1:27 – 28)

While part of Manasseh settled in the land of Gilead (see below), the other part settled in Canaan. Only the southwestern boundary is described in some detail (Josh 17:7 – 11). Manasseh stretched from the Mediterranean Sea to the Jordan River; the southern boundary was conterminous with Ephraim, while on the north Manasseh bordered on Asher,

Zebulun, and Issachar. (map p. 56) Cities such as Dor, Megiddo, Beth Shan, and Taanach were difficult to capture because of the strength of the Canaanites in the plains. Thus, Manasseh settled in the heavily forested hill country, cutting down trees to secure needed farmland (vv. 15–18).

Next, Joshua 18:1–10 focuses on Shiloh in the Hill Country of Ephraim, where the tabernacle was set up. From there, Joshua sent out three men from each of the remaining tribes to prepare a written survey of the remaining portions of Canaan. Based on this survey, the remaining allotments were made.

Benjamin (Josh 18:11–28; Judg 1:21)

Benjamin's territory fell between the two powerful tribes of Ephraim (to the north) and Judah (to the south). Of particular interest Jerusalem ("the Jebusite city") was located in Benjamin, not in Judah (Josh 18:16–17).

The strategic importance of Benjamin cannot be overemphasized. One of the main approach roads from the coastal plain into the hill country ran through its western portion. On the east, several roads led down into the rift valley and joined at the oasis of Jericho and from there proceeded across the fords of the Jordan into Transjordan. Thus, Benjamin was one of the busiest tribal areas, for invading international powers often entered the hills via the roads from the east or the west, and the northern and southern Israelite kingdoms occasionally met in battle in the territory of Benjamin as they sought to expand their influence.

Zebulun (Josh 19:10–16; Judg 1:30)

The greater part of Zebulun's allotment was confined to the high ground overlooking the Jezreel Valley to the south; on occasion, its territory stretched across the valley to include a city such as Jokneam (cf. Josh 21:34).

Issachar (Josh 19:17 – 23)

The towns given to Issachar were located in the valleys and on the basalt heights of eastern Lower Galilee as well as in the eastern portion of the Jezreel Valley. Because of the basalt rocks and the lack of water sources, the heights were never densely settled; the major settlements were located in the valleys. The major international highways ran close to its southern and western borders, and it would have been easy for armies on the march to invade.

Asher (Josh 19:24 – 31; Judg 1:31)

Asher was apportioned territory in the northwestern corner of Israel. Its territory stretched from Mount Carmel in the south to the Litani River on the north and theoretically included the coastal area. But Asher was unable to take control of important cities located on the coastal plain (Judg. 1:31). The one natural harbor of the whole country, Acco, was located in Asher's allotment, but it was rarely controlled by Israel.

Naphtali (Josh 19:32 – 39; Judg 1:33)

The prominent hill of Mount Tabor served as the meeting point for the boundaries of three tribes — Issachar, Naphtali, and Zebulun. Since much of Galilee was unoccupied prior to the Israelites' arrival, Naphtali was probably able to settle in the hills relatively easily.

Dan (Josh 19:40 – 48; Judg 1:34 – 35; 17 – 18)

Dan's eastern boundary, on the western slopes of the mountains overlooking the coastal plain, coincided with the western boundary of Benjamin. Its southern boundary was identical to Judah's northern one and followed the Sorek Valley to the Mediterranean Sea (Josh 15:10 – 11).

Dan's territory sat astride the Aijalon Valley, through which the major approach road into the Hill Country of Ephraim, Benjamin, and Judah ran. Moreover, the main international north – south route ran through

TRIBES NORTH OF THE JEZREEL VALLEY: ZEBULUN, ISSACHAR, ASHER, AND NAPHTALI

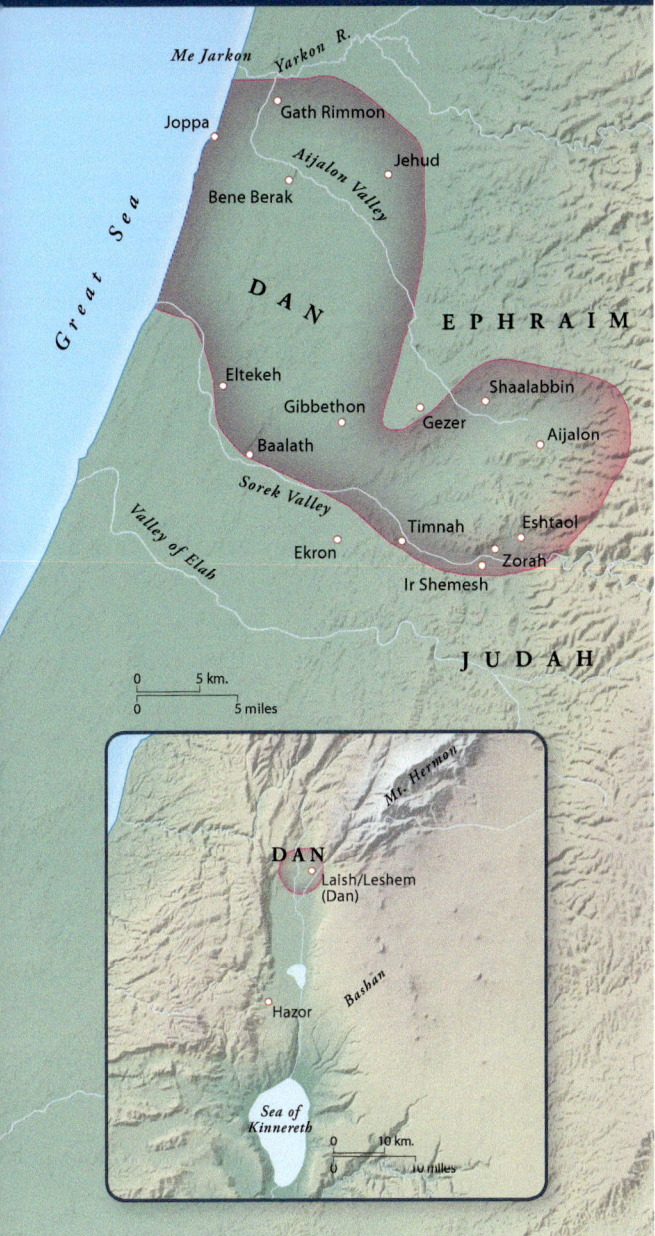

Me Jarkon · Yarkon R. · Joppa · Gath Rimmon · Jehud · Aijalon Valley · Bene Berak · Great Sea · D A N · E P H R A I M · Eltekeh · Shaalabbin · Gibbethon · Gezer · Aijalon · Baalath · Sorek Valley · Timnah · Eshtaol · Ekron · Zorah · Ir Shemesh · Valley of Elah · J U D A H

0 5 km. / 0 5 miles

Mt. Hermon · D A N · Laish/Leshem (Dan) · Bashan · Hazor · Sea of Kinnereth · 0 10 km. / 0 10 miles

Mt. Hermon · M A N A S S E H · Bashan · Sea of Kinnereth · Ashtaroth · Settlements of Jair · Edrei · Debir · Ramoth Gilead · Jordan R. · G A D · G i l e a d · Zaphon · Succoth · Jabbok R. · Mahanaim · Ramath Mizpah · Jogbehah · A M M O N I T E S · Betonim · Rabbah · Beth Nimrah · Jazer · Beth Haram · Heshbon · Mephaath · Beth Jeshimoth · Beth Peor · Kiriathaim · Medeba · Beth Baal Meon · Mishor · Zereth Shahar · Jahaz · R E U B E N · Ataroth · Dibon · Kedemoth · Aroer · Salt Sea · Arnon Gorge · M O A B I T E S · 0 10 km. / 0 10 miles

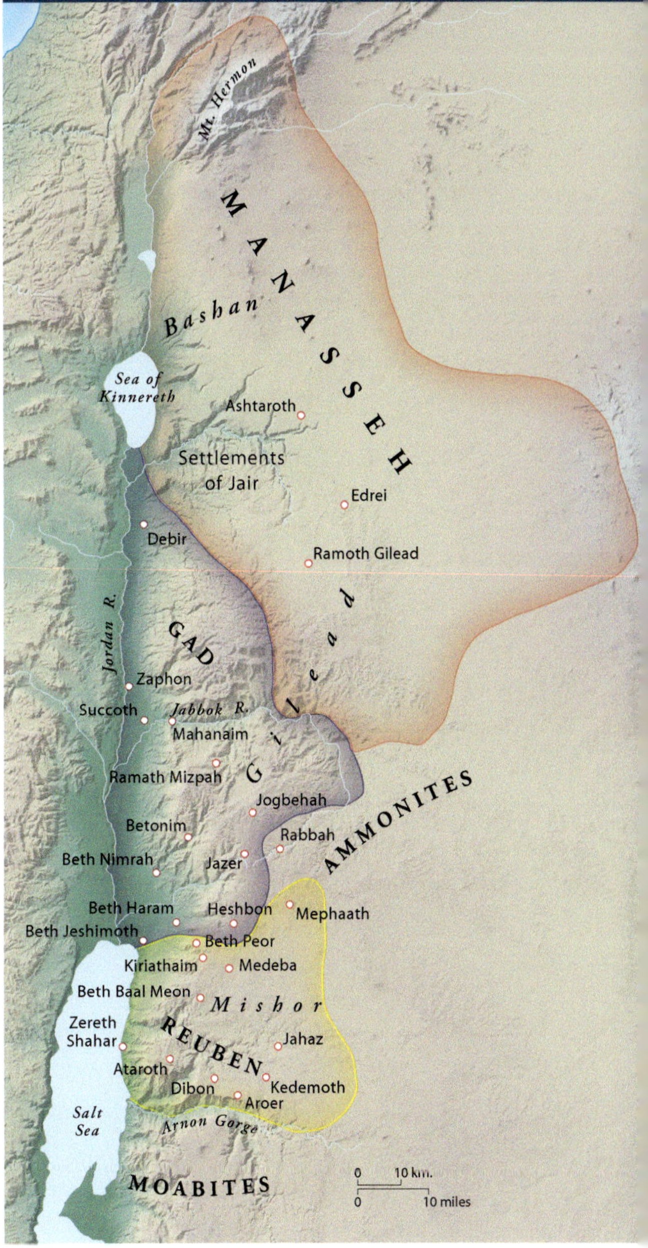

Dan's western extension. Therefore, the Danites were unable to expand westward but were confined to the western slopes of the mountains (see Judg 1:34 – 35; 13 – 16). Due to Amorite pressures, some of the Danites moved northward to Laish/Leshem, which they captured and renamed "Dan" (Judg 17 – 18).

Reuben, Gad, and Manasseh (Josh 13:8 – 33; Num 32)

Reuben's territory (Josh 13:15 – 23; Num 32:37 – 38) stretched from the Arnon Gorge in the south to the city of Heshbon in the north. The territory allotted to Gad stretched from Heshbon to

LEVITICAL CITIES ALLOTTED WITHIN THE TRIBAL AREAS, AND CITIES OF REFUGE

Great Sea (Mediterranean Sea)

Abdon
Acco
Rehob
Mishal
Rimmon
Helkath
Jokneam
Taanach
Kartan
En Gannim
Daberath
Kishion
Jarmuth

ASHER
NAPHTALI
ZEBULUN
ISSACHAR

Kedesh

Sea of Kinnereth

MANASSEH

Golan
Ashtaroth

Yarmuk R.

Ramoth Gilead

MANASSEH

Jordan R.

Shechem

Gath Rimmon
Joppa

EPHRAIM
Shiloh

Beth Horon
Eltekeh
Gibbethon
Gezer
Aijalon
Gibeon
Beth Shemesh

DAN

BENJAMIN
Geba
Almon
Anathoth
Jerusalem

Jabbok R.
Mahanaim

GAD

Jazer
Mephaath

Heshbon

Libnah

JUDAH

Hebron

Debir
Juttah
Jattir
Eshtemoa

Beersheba

SIMEON

REUBEN
Bezer
Jahaz
Kedemoth

Salt Sea (Dead Sea)

Arnon Gorge

Zered River

Gaza

Hebron	Cities of refuge (underlined)

Levitical cities:

△ Towns received by Kohathite clans
△ Towns received by Gershonite clans
△ Towns received by Merarite clans

0   10 km.
0   10 miles

Ahlab

Tyre

Beth
Anath

Dan

*Upper
Galilee*

MAACAHITES

Aczib

Beth
Shemesh

SIDONIANS

Acco

Rehob

GESHURITES

Aphek

*Lower
Galilee*

*Sea of
Kinnereth*

CANAANITE/EGYPTIAN

Yarmuk R.

Dor

Megiddo

Taanach

Beth Shan

Ibleam

*G r e a t   S e a
(Mediterranean Sea)*

Jordan R.

*Hill Country
of Manasseh*

Shechem

Jabbok R.

G i l e a d

*Hill Country
of Ephraim*

C o a s t a l   P l a i n

Shaalbim

Gezer

Aijalon

Rabbah
of the Ammonites

Ekron

Jebus

Ashdod

Beth
Shemesh

JEBUSITES

*Tableland
of Moab
(Mishor)*

Ashkelon

Gath

CANAANITE
(LATER PHILISTINE)

Gaza

Hebron

*Hill Country
of Judah*

*Salt Sea
(Dead
Sea)*

Arnon Gorge

AVVITES

Zered R.

	Limits of Israelite settlement
Gath	Unconquered Canaanite centers (bold type)

0   10 km.

0   10 miles

SHAMGAR

Hazor (Jabin)

ELON

Sea of
Kinnereth

Kishon R.

Kedesh
(of
Naphtali)
(Barak)

JAIR

Mt.
Tabor ▲

Havvoth
Jair

GIDEON

▲ Mt. Moreh

Megiddo

En Harod

Kamon

Ophrah

Taanach

Jabesh
Gilead

Abel
Meholah

TOLA

JEPHTHAH

Jordan R.

Shamir

Mt. Ebal ▲

Zaphon

ABDON

Penuel

Pirathon

▲ Shechem

Succoth

Jabbok R.

Mt.
Gerizim

Mizpah

Great Sea

Shiloh

Gilead

DEBORAH

Bethel

EHUD

Mizpah

Gilgal

Ramah

Rabbah
(of the Ammonites)

Sorek Valley

Timnah

Gibeah

City of Palms
(Jericho)

Eshtaol

Zorah

Jerusalem

IBZAN

SAMSON

Bethlehem

AMMONITES

PHILISTINES

Ashkelon

Tableland
of Moab
(Mishor)

Gaza

Hebron

OTHNIEL

Debir

Salt
Sea

Arnon Gorge

Beersheba

| GIDEON | Major judge |
| ELON | Minor judge |

0   10 km.

0   10 miles

Mahanaim on the Jabbok River in the north (map p. 57) and from just west of Rabbah (of the Ammonites) to the Jordan River. Gad also seems to have received the whole Jordan Valley east of the river (Josh 13:27 – 28). Certain clans from the tribe of Manasseh settled north of the Jabbok.

Throughout the rest of Old Testament history these tribes felt continuous pressure from the Moabites, Ammonites, Midianites, Ishmaelites, and the Arameans of Damascus. Despite these pressures an Israelite/Jewish presence was maintained in Transjordan through the end of the biblical period and beyond.

Levitical Cities and Cities of Refuge (Josh 20 – 21)

Joshua and Eleazar assigned towns and their pasturelands to the Levites (map p. 58). These forty-eight cities were assigned to the three Levitical clans, and approximately four such cities were designated from each of the tribes. Thus in their towns, scattered throughout Israel, the Levites probably were spreading a godly influence through their teaching activities and eventually loyalty to the Davidic dynasty. Many Levites moved south to Judah and Jerusalem when Jeroboam I revolted against Solomon's son, Rehoboam (931 BC; 2 Chron 11:13 – 17).

Six cities were designated as places to which a person who committed manslaughter could flee. There the case would be tried (Josh 20:4, 6), and if it was judged as accidental manslaughter rather than murder, the slayer was required to remain in that city until the death of the high priest (Num 35:9 – 34; Deut 4:41 – 43; 19:1 – 14; Josh 20:6). Three of the cities were located to the west of the Jordan River and three to the east.

The Period of the Judges

In theory at least, the land God had promised to Abraham, Isaac, and Jacob had been assigned to their descendants. But the Israelites were well aware that not all of the land had been conquered (map p. 59). The southwestern portion was counted as Canaanite and would later become Philistine (Josh 3:2 – 3; Judg 3:3). The Jezreel Valley remained in Canaanite and/or Egyptian hands (Judg 1:27 – 28), while the Mediterranean coastline north of Mount Carmel was under Sidonian control (Josh 13:4; Judg 1:31 – 32; 3:3). Certain foreign enclaves

remained independent, such as the city of Jebus (Jerusalem; Josh 15:63; Judg 1:21). Most of these areas came under Israelite control by the days of David and Solomon (see 1 Kings 4:7 – 19).

The process of Israelite settlement must have proceeded in a somewhat peaceful fashion as farmsteads were established, scrub forests cleared, terraces built, and crops (especially grapes, olives, figs, almonds, and wheat) planted. In addition, the recent invention of rock-hewn, plaster-lined cisterns allowed the people to settle in areas distant from springs. During this period Israel was able to remain in the mountains, somewhat removed from the threats of Egyptian kings, who were concerned with controlling the international routes.

From Egypt, almost 400 cuneiform tablets were discovered at el-Amarna. Most of the letters are records of the correspondence between the numerous city-state rulers in the Levant and the king of Egypt. One of the most prominent of these rulers was Labayu of Shechem. From his base in the hill country, he reached out in all directions attempting to take control of the trade routes that ran through the country, harassing other rulers. Eventually Labayu's enemies in Canaan murdered him, though his sons soon followed in his footsteps.

In the process of settling down, some Israelites began to worship pagan deities, including Baal (responsible for the fertility of the land) and his consort Ashtoreth (goddess of war and fertility), as well as Asherah. As punishment for these sins, Yahweh sent foreigners who oppressed the Israelites. Eventually Israel cried out in repentance, and God sent "judges" to deliver Israel from her oppressors and to usher in periods of "rest." This cycle of lapse into sin, divine punishment, repentance, deliverance, and rest is the story of Judges (Judg 2:10 – 19).

The Judges Othniel and Ehud (Judg 3:7 – 30)

The first major judge was Othniel, who delivered the Israelites from the hands of Cushan-Rishathaim. It is somewhat uncertain where his home territory (Aram Naharaim) was located.

Next was Ehud, a Benjamite (Judg 3:15). Eglon, the king of Moab, along with his Transjordanian neighbors, oppressed Israel from their headquarters in the "City of Palms" (Jericho, according to 2 Chron 28:15). Ehud, after delivering tribute to Eglon, killed him and drove the Moabites back across the

THE SEA PEOPLES

DORIAN GREEKS

ASIA MINOR (ANATOLIA)

Aegean Sea

Mediterranean Sea

CRETE

CYPRUS

Ugarit

CANAAN

SEA PEOPLES: Tjekker, Danuna, Sherden, Tursha, Lukka, Sheklesh and Philistines

EGYPT

0   100 km.

0   100 miles

and of the prophetess Deborah, the Israelites mustered their forces on the slopes of Mount Tabor at the northeastern end of the Jezreel Valley (map p. 60) while the Canaanites gathered near Megiddo in the Jezreel Valley. On the wooded slopes of Mount Tabor the Israelites were relatively safe from the operations of the nine hundred chariots controlled by Sisera (4:3 – 13).

Barak led his forces down the mountain to battle the Canaanites in the plain below. Because of a sudden downpour and the flooding of the Kishon River, the Canaanite chariots were rendered ineffective, and Israel scored a decisive victory. Sisera, fleeing back toward Hazor, was killed by the wily Jael, who drove a tent peg through his head while he slept (4:18 – 21). The northern portion of the land experienced relative peace for forty years (5:31).

Jordan. At least a portion of the land then enjoyed "peace" for eighty years.

Although the Bible does not mention it, the Egyptians were active in Canaan at the end of the fourteenth and in the thirteenth centuries BC. For example, Seti I suppressed a revolt in the Beth Shan area and reasserted Egyptian control there. His successor, Rameses II (1279 – 1213 BC), battled the Hittites at Kadesh on the Orontes in his fifth year. Merneptah (1213 – 1203 BC), the successor of Rameses II, was not as powerful as his forerunner, but his stele, among other items, mentions a victory over Israel: "Israel is laid waste, his seed is not." This is the first extrabiblical reference to "Israel" in Canaan.

Why does the Bible not mention this Egyptian activity going on in Canaan at this time? Most likely the Egyptians were interested in the plains and open valleys, where the international routes ran, while the Israelites were striving to secure a foothold in the more rugged mountainous areas.

Deborah and Barak (Judg 4 – 5)

During the first half of the twelfth century BC, the Canaanites were again asserting themselves in northern Israel. Led by Jabin, the king of Hazor, and his commander Sisera, they oppressed northern Israel for twenty years. Under the leadership of Barak

Gideon (Judg 6 – 8)

Next, the Midianites and their allies invaded Israel in the late spring/early summer, confiscating the newly harvested crops and grazing their livestock and camels in the fields. Gideon, from the tribe of Manasseh and the village of Ophrah (map p. 60), which overlooked the Jezreel Valley, responded to God's call. Mustering his troops at the spring of Harod, Gideon prepared for battle by culling his troops down to a select 300. Dividing this corps into three units, he surprised the Midianites — whose main camp was evidently on the northern slope of Mount Moreh at Endor (Ps 83:10) — with a night attack. The frightened Midianites fled in a southeasterly direction toward the Jordan River.

The Ephraimites joined the battle by seizing control of the fords of the Jordan. Gideon and his troops pursued the Midianites eastward, passing Succoth and Peniel (Judg 8:8 – 9) and heading up the Jabbok River into Gilead, and defeated them.

Jephthah (Judg 10:6 – 12:7)

At the beginning of the eleventh century BC, the Israelites were pressed on the east by the Ammonites, especially in Gilead (map p. 60). The major dispute was over the control of the tableland of Moab and southern Gilead. The elders of Gilead chose Jephthah to head their forces. The ensuing battle with the Ammonites most likely took place south of Mizpah of Gilead (11:29 – 33), and the Ammonites were soundly defeated.

Samson (Judg 13 – 16)

Around 1200 BC a group of tribes, known collectively as the "Sea Peoples," made their way into the lands of the eastern Mediterranean. One of the tribes, the Philistines, settled along the southern coast of Canaan in Gaza, Ashkelon, Ashdod, Gath, and Ekron. Their early adoption of new technology for forging metal products gave them a military edge, and they quickly assimilated major features of Canaanite culture, including its religion (note Canaanite deities such as Dagon, Ashtoreth, and Baal-Zebub), pottery, and probably a Semitic language.

During the first half of the eleventh century BC the Philistines began to exert pressure on the Israelites. The natural place for this conflict was in the buffer zone between them, the Shephelah. There, in the Valley of Sorek, the Nazirite judge Samson rose to meet the threat (map p. 60). His earliest encounter took place when he married a Philistine woman who lived in Timnah, located in the Sorek Valley. Later, as an offended husband, he burned the fields of grain, the vineyards, and the olive groves that the Philistines maintained in the Sorek Valley. Although his exploits had taken him to Ashkelon (14:19) and would eventually lead him to Gaza and Hebron (16:1 – 3), it was an affair with "a woman in the Valley of Sorek whose name was Delilah" that led to Samson's heroic but tragic death at Gaza (16:4 – 31).

With the death of Samson, the last of the heroes of the book of Judges passed from the scene. Yet even at the end of this period large pockets of non-Israelite populations still existed in the country. King Saul would provide protection from some of these enemies, but it was King David who eventually brought them to submission.

ARRIVAL AND SETTLEMENT OF THE PHILISTINES

Mediterranean Sea

Tyre
Dan
Hazor
Sea of Kinnereth
Dor
Tel Qiri
Afula
Megiddo
Taanach
Beth Shan
TJEKKER
Jordan R.
Tel Qasile
Tel Aphek
Joppa
Azor
Bethel
Gezer
Tel Masos
Ashdod
Timnah
Ekron
Jerusalem
Beth Shemesh
Ashkelon
Gath
Beth Zur
Lachish
Gaza
Tel Eton
Dead Sea
PHILISTINES
Beersheba
Tel Masos

0   10 km.
0   10 miles

○ Some sites with Aegean-type artifacts

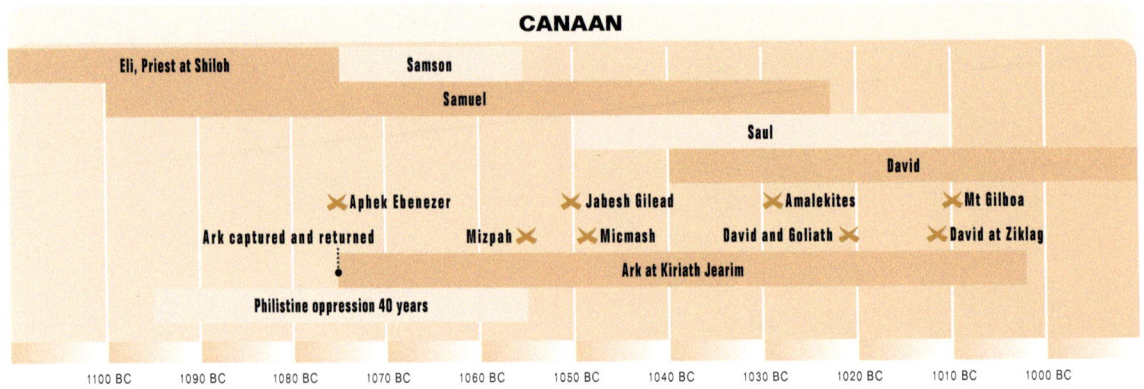

CANAAN

Eli, Priest at Shiloh

Samson

Samuel

Saul

David

✗ Aphek Ebenezer   ✗ Jabesh Gilead   ✗ Amalekites   ✗ Mt Gilboa

Ark captured and returned   Mizpah ✗   ✗ Micmash   David and Goliath ✗   ✗ David at Ziklag

Ark at Kiriath Jearim

Philistine oppression 40 years

1100 BC   1090 BC   1080 BC   1070 BC   1060 BC   1050 BC   1040 BC   1030 BC   1020 BC   1010 BC   1000 BC

TRANSITION TO THE MONARCHY: SAMUEL AND SAUL

At the beginning of the eleventh century Samuel was born to parents who lived in Ramah. During his childhood years he served the high priest Eli at the tabernacle at Shiloh (1 Sam 3). The tabernacle was probably located there because the rugged terrain of the Hill Country of Ephraim provided natural topographical defenses.

When Samuel was approximately twenty-five years old, the Philistine forces gathered at Aphek to invade the hill country (1 Sam 4:1). To meet this threat, the Israelites set up camp near Ebenezer. After an initial defeat, the Israelites believed that if the visual symbol of God's presence (the ark) would go with them into battle, they would be victorious. But even after the ark was brought from Shiloh to Ebenezer, the Philistines soundly defeated the Israelites and even captured the sacred ark! Archaeological evidence suggests that Shiloh was destroyed at this time.

The ark was first taken to Ashdod and placed in the temple of Dagon. But the statue of Dagon repeatedly fell over in obeisance to the ark, and a plague broke out in the city, so the ark was sent to Gath (5:1 – 8). There a similar plague broke out,

and the troublesome ark was transferred to Ekron, where a similar plague broke out. Anxious to rid themselves of the ark, the Philistines placed it on a cart pulled by cows, who headed eastward up the valley to Beth Shemesh. Shortly thereafter the ark was transferred to Kiriath Jearim, a Gibeonite city, where it remained until David transported it to Jerusalem (2 Sam 6).

Samuel lived in Ramah, strategically located at the important junction of the west – east connecting route (Gezer – Beth Horon – Ramah – Jericho) and the north – south Ridge Route. His annual duties took him on a circuit that led from Ramah to Bethel, to Gilgal, and to Mizpah (1 Sam 7:16).

Philistine pressure on Israel continued, and the ensuing battle occurred somewhere on the western side of the Benjamin Plateau (1 Sam 7). God intervened on Israel's behalf, and Israel pursued the Philistines westward toward the coastal plain. Israel regained territory in the Ekron and Gath areas (vv. 11 – 14), but the lull in the Philistine threat lasted only a few years.

For a variety of reasons the elders of Israel asked Samuel to appoint a king over them. At this point in the narrative (1 Sam 9), Saul of Benjamin is introduced. Looking for lost donkeys in the Hill Country of Ephraim, Saul approached Samuel for guidance. In turn Samuel privately anointed him king (10:1). Soon after, Israel assembled at Mizpah, where Saul was chosen by lot to be king (vv. 17 – 27). Saul then returned to his residence in Gibeah (Tell el-Ful), which he renamed Gibeah of Saul and from where he ruled Israel.

Saul soon had the opportunity to exhibit his leadership qualities by mustering Israelite and Judean forces for the purpose of delivering the people of Jabesh Gilead, across the Jordan, from their Ammonite oppressors (1 Sam 11:1–13). Israel then confirmed him as king at the old cultic center of Gilgal (vv. 14–15).

Next, Saul and his son Jonathan mustered small numbers of Israelite forces in Micmash and Gibeah of Benjamin (1 Sam 13:2). Most likely the Philistines controlled western Benjamin by means of garrisons at Gibeah of God (10:5) and Geba/Gibeon (13:3). Saul's son Jonathan met the Philistine threat head-on by assaulting their garrison at Geba/Gibeon. The Philistines regrouped and reentered the hill country from the north, which brought them back into the Benjamin plateau. They brought with them chariots, horsemen, and foot soldiers and set up camp at Micmash (v. 5).

After assembling troops at Gilgal, Saul and Jonathan moved into the hill country. At Micmash, Jonathan and his armor-bearer subdued the guards of the Philistine camp and the panicked Philistines fled (1 Sam 14:14–15). Saul led the remaining Israelite forces into the fray and was able to drive the Philistines out of the hill country back to the Aijalon region (v. 31).

After a battle with the Amalekites Samuel anointed David (1 Sam 16), and David began to serve at Saul's court as a musician. The Philistines and

SAMUEL, THE PHILISTINES, AND THE CALL OF SAUL

Legend:
- Passage of the ark during battles with Philistines
- Towns on Samuel's judicial and priestly circuit

0   10 km.
0   10 miles

SAUL'S BATTLES AGAINST ISRAEL'S ENEMIES

Legend:
- Saul's forces
- Philistine forces
- Philistine raiding parties

0   5 km.
0   5 miles

0   2 km.
0   2 miles

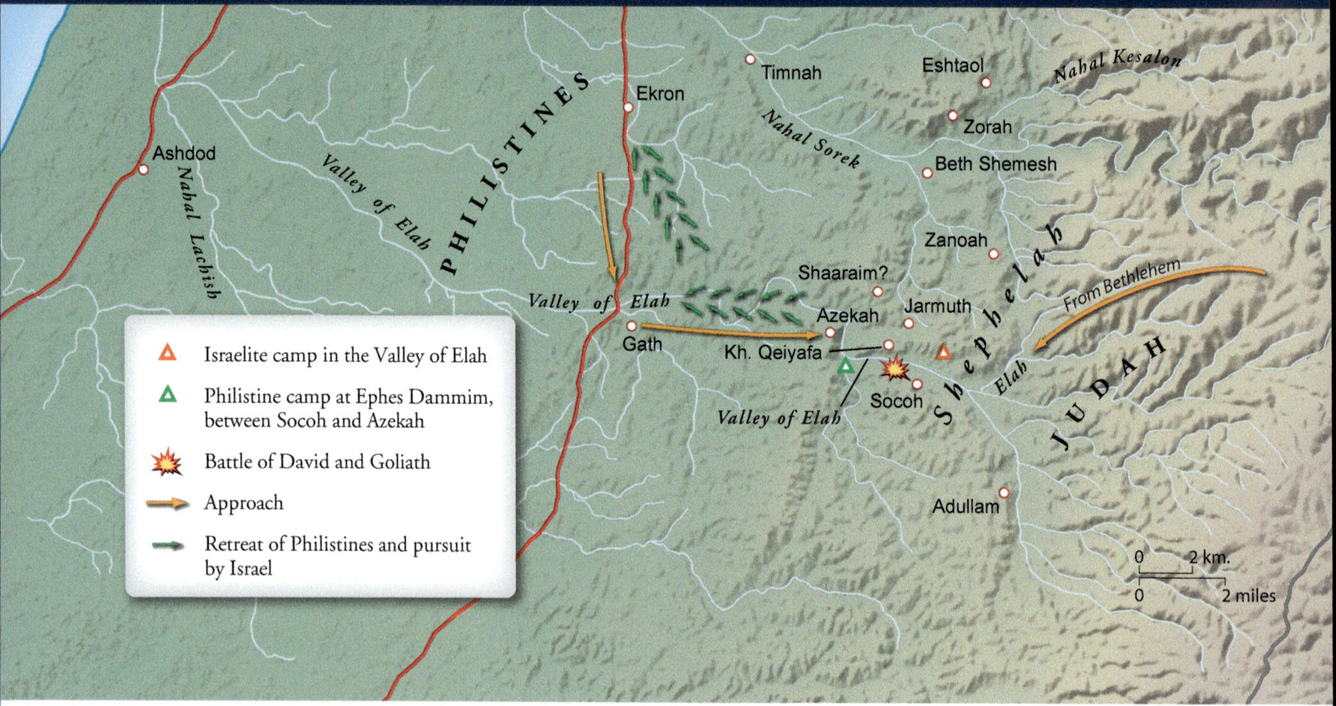

Ashdod

Nahal Lachish

Valley of Elah

PHILISTINES

Ekron

Timnah

Eshtaol

Nahal Kesalon

Zorah

Nahal Sorek

Beth Shemesh

Zanoah

Shaaraim?

Jarmuth

Azekah

Gath

Kh. Qeiyafa

Valley of Elah

Socoh

Valley of Elah

Shephelah

From Bethlehem

Elah

JUDAH

Adullam

- 🔺 Israelite camp in the Valley of Elah
- 🔺 Philistine camp at Ephes Dammim, between Socoh and Azekah
- 💥 Battle of David and Goliath
- ➡ Approach
- ➡ Retreat of Philistines and pursuit by Israel

0   2 km.

0   2 miles

🔺 *Valley of Elah from Kh. Qeiyafa. Looking west with Azekah on the left (south) side of the image. David fought Goliath near here (1 Sam 17).*

Israelites were still vying for power, but now their battles were fought in the Shephelah, a military buffer zone. The Philistines moved into the Valley of Elah (1 Sam 17). The Israelites camped on the north side of the valley (vv. 2 – 3), probably east of the Philistine camp.

Here in the Valley of Elah David killed Goliath. Emboldened by David's example, Saul's troops successfully attacked the Philistines, who fled to the security of Gath and Ekron (v. 52). Following this stunning victory, 1 Samuel 18 – 21 describes David's growing popularity among the people and Saul's growing jealousy. These chapters are punctuated with descriptions of David's narrow escapes from Saul, most of the action occurring in the Benjamin area. David eventually fled Saul's court but was pursued throughout Judah (1 Sam 22 – 26), one of the prominent places being En Gedi on the shores of the Dead Sea.

David must have realized that he would either have to kill Saul in self-defense or else be killed by him. To avoid this, David again sought asylum with Achish, the king of Gath. By now Achish was well aware that David was a true enemy of the Israelite king. Achish, planning to make use of David's troops and military prowess, stationed him at Ziklag. There, on his southern border, David was assigned to protect Achish against raiders from the south.

DAVID AND SAUL

Legend:
- David's flight from Saul
- Philistines
- Passage of Saul's body after his death
- David's journeys after Saul's death

- (a) David receives Goliath's sword
- (b) Achish, Philistine king of Gath
- (c) David assembles fighting force
- (d) David's parents transferred to Moab
- (e) David delivers Keilah from Philistines
- (f) David flees from Saul
- (g) David hides from Saul in a cave
- (h) Home of Nabal and Abigail
- (i) David defends southern border of Philistines (and Judeans)
- (j) David pursues Amalekites
- (k) Saul and three of his sons fall at Mt. Gilboa

▲ *Oasis of En Gedi on the western shore of the Dead Sea. Here David hid from Saul (1 Sam 24).*

▲ *David worked for the Philistines in this area of the western Negev/ Besor Ravine (1 Sam 27; 30). The light soil is "loess soil."*

David did, in fact, conduct raids on the Geshurites, the Girzites, and the Amalekites, all of whom dwelt in northern Sinai between Ziklag and Shur (1 Sam 27:8). Leaving no survivors, David was able to perpetrate a lie by telling Achish, who lived in Gath, that he had made raids on the Negev of Judah, the Negev of Jerahmeel, and the Negev of the Kenites (27:10). In this way David was leading Achish to believe that Judean hostility toward David was growing, when in fact Judean appreciation for David was increasing because he was defending them from these desert bands!

After David had served sixteen months at Ziklag (1 Sam 27:7), the final confrontation between Saul and the Philistines began to unfold, though we are not exactly sure how.

In any case, the Philistines assembled their forces at Aphek (map p. 67), and from there they marched farther north, setting up camp opposite Israel on the southern slope of Mount Moreh at Shunem in the Jezreel and Harod Valley region.

There the Israelites were engaged in a life-and-death struggle with the Philistines. Some of the fleeing Israelites went up Mount Gilboa, hoping the mountain would offer protection from the pursuing Philistine chariots. There on Mount Gilboa Saul and Jonathan died (1 Sam 31). With the burial of Saul and Jonathan by the men of Jabesh Gilead, the transition period between the period of the judges and that of the monarchy ended. Within a few years the idea of a dynastic monarchy would be firmly established, at least in the minds of the Judeans.

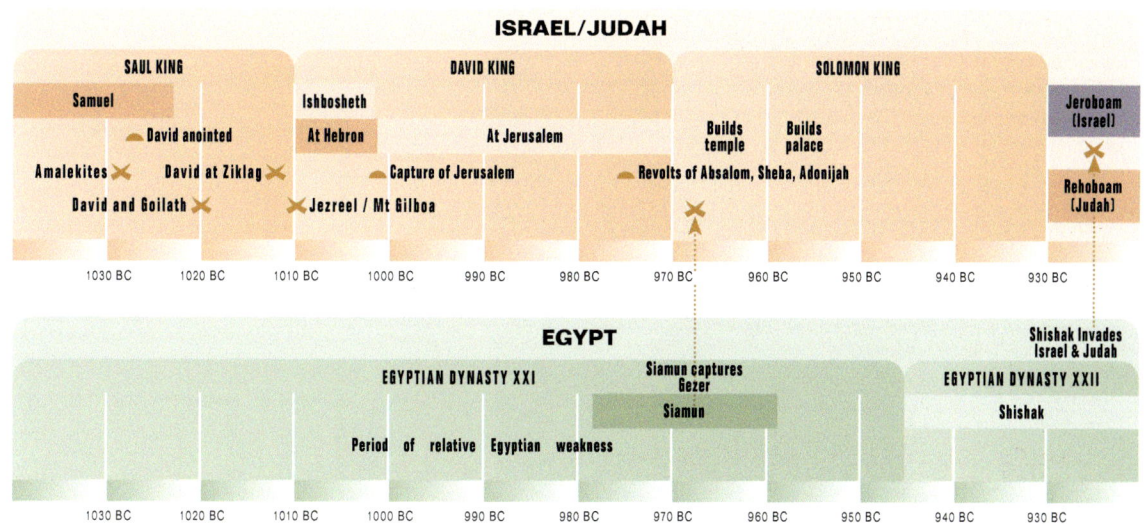

ISRAEL/JUDAH

SAUL KING — Samuel — David anointed — Amalekites — David at Ziklag — David and Goliath

DAVID KING — Ishbosheth — At Hebron — At Jerusalem — Capture of Jerusalem — Jezreel / Mt Gilboa

SOLOMON KING — Builds temple — Builds palace — Revolts of Absalom, Sheba, Adonijah

Jeroboam (Israel) — Rehoboam (Judah)

1030 BC 1020 BC 1010 BC 1000 BC 990 BC 980 BC 970 BC 960 BC 950 BC 940 BC 930 BC

EGYPT

EGYPTIAN DYNASTY XXI — Period of relative Egyptian weakness

Siamun captures Gezer — Siamun

EGYPTIAN DYNASTY XXII — Shishak — Shishak Invades Israel & Judah

1030 BC 1020 BC 1010 BC 1000 BC 990 BC 980 BC 970 BC 960 BC 950 BC 940 BC 930 BC

THE UNITED MONARCHY: DAVID AND SOLOMON

▼ Gihon Spring: The spring chamber of the chief water source of Jerusalem

David as King

After Saul died, one of his surviving sons, Ish-Bosheth, became king (2 Sam 2:9) and ruled from the Transjordanian city of Mahanaim. David moved to Hebron, where he was anointed king of Judah (2:1 – 7, 11). After Ish-Bosheth's murder two years later (4:1 – 12), the elders of Israel came to David at Hebron, made a covenant with him, and anointed him king over all Israel (5:1 – 3).

The Philistines realized a united Israel posed a serious threat to their control of the country. They responded by twice trying to divide the country into two parts and establishing themselves in the hill country in the Valley of Rephaim, southwest of Jerusalem, but David drove them back to the coastal plain (2 Sam 5:17 – 25; 1 Chron 14:8 – 17).

Labels on map: HAMATH; Cun; Gebal; Lebo Hamath; ZOBAH; Berothai; BETH REHOB; Sidon; Abel Beth Maacah; Damascus; ARAMEANS; Tyre; Dan; MAACAH; Mediterranean Sea; Aczib; Hazor; GESHUR; Acco; Geshur; Plain of Acco; GALILEE; Sea of Galilee; Helam; Jezreel Valley; TOB; Megiddo; Harod Valley; Ramoth Gilead; Taanach; Beth Shan; Tob; Ibleam; Jabesh Gilead; Jordan R.; Shechem; Mahanaim; Joppa; Upper Beth Horon; Rabbah of the Ammonites; Gezer; David makes Jebus his political and religious capital; AMMON; Baalah of Judah/ Kiriath Jearim; Gibeon; Jericho; Jerusalem; Medeba; Gath; Baal Perazim; Joab battles Ammonites and allies from the north; Gaza; Hebron; Dead Sea; Aroer; Ziklag; Arnon Gorge; Beersheba; Kir Moab; MOAB; Valley of Salt; AMALEKITES; EDOM

0   20 km.
0   20 miles

Israelites
Arameans
Edomites
Ammonites
Subdued by David

that Jerusalem became the religious as well as the political center. From a defensive standpoint its mountainous location meant that any enemies wishing to attack Jerusalem had to make the difficult ascent from the coastal plain. Moreover, its local topography made it easily defensible on the western, southern, and eastern sides.

David consolidated his kingdom internally by taking control of the old Canaanite centers located in the Jezreel and Harod Valleys — Megiddo, Taanach, Ibleam, and Beth Shan (Judg 1:21 – 35). To the southwest he subdued the Philistines (1 Chron 18:1).

David's first war in the Transjordanian region was with the Ammonites, who hired Aramean troops from the north to help them (1 Chron 19:6 – 7). Joab and his brother Abishai drove off the Aramean and Ammonite forces (2 Sam 10:6 – 14; 1 Chron 19:6 – 15), and Israel pressed its advantage by conquering the territory of Moab (2 Sam 8:2; 1 Chron 18:2).

The Arameans regrouped at Helam, but David and his troops won the battle (2 Sam 10:15 – 19). The Israelite forces then returned to Rabbah to deal decisively with the Ammonites. During the siege of Rabbah David was back in Jerusalem, having his sinful affair with Bathsheba (2 Sam 11). After David repented of his sins (12:1 – 24), Rabbah was captured (12:26 – 31).

Finally, David's army defeated the Edomites in the Valley of Salt (2 Sam 8:13 – 14; 1 Chron 18:12 – 13). Hadad, of royal Edomite descent, fled to Egypt via Midian and Paran and later returned to lead his people in revolt against Solomon (1 Kings 11:14 – 22).

David now controlled many of the neighboring states. To the far north the king of Hamath acknowledged David's

After ruling in Hebron for seven years, David captured Jebus (Jerusalem; 2 Sam 5:6 – 10). David made this non-Israelite city his personal possession and that of his descendants, so that neither Judah nor Israel could lay claim to it. He also brought the ark of the covenant to Jerusalem (2 Sam 6; 1 Chron 13), so

supremacy by sending his son along with "gifts" of silver, gold, and bronze to Jerusalem. Major Aramean areas came under Israelite control; a garrison was even placed in Damascus (2 Sam 8:5–6; 1 Chron 18:5–6). Talmai, the king of Geshur, formed an alliance with David and sealed it by marrying his daughter to David; the offspring of their union was Absalom (2 Sam 3:3). Absalom fled there after killing his half-brother Amnon, who had raped Absalom's sister Tamar (13:37–39).

Although the exact boundaries of the David's empire are not given, he seems to have controlled northern Sinai all the way to the eastern Nile delta (see 1 Chron 13:5) and in the far north up to Lebo Hamath. The Israelites did now, in fact, rule over most of the land of Canaan promised them as an inheritance 400 years earlier (Num 34; map p. 47). By David's day, the traditional boundaries "from Dan to Beersheba" were firmly established (2 Sam. 24:2).

Toward the end of David's reign the question of his successor became acute. His oldest living son, Absalom, decided to take matters into his own hands (2 Sam 14–19). After having been proclaimed king at Hebron, Absalom with his followers marched north to Jerusalem. David and his followers fled eastward from Jerusalem, crossed the Jordan, and found refuge in Mahanaim. In the "forest of Ephraim" (18:6), Joab defeated and killed Absalom.

Upon David's return to Jerusalem he had to deal with a second revolt, this one led by Sheba (2 Sam 20:1–22). As usual, it was Joab who dealt decisively with the matter, pursuing Sheba to the northern Israelite city of Abel Beth Maacah, where Sheba was killed.

When David became old, his (apparently) eldest son, Adonijah, decided to secure the kingship for himself (1 Kings 1:5–27). With the assistance of Joab, the commander of the army, and Abiathar the priest, he assembled many of the notables of the kingdom near En Rogel, a small spring located in the Kidron Valley just south of Jerusalem (vv. 5–10). Upon hearing of this, Nathan the prophet and Bathsheba approached David in order to confirm Solomon as king, and they did so at the Gihon Spring (vv. 11–48). After David died, Solomon inherited kingdom.

Solomon as King

Most of Solomon's rule was taken up with building projects and commercial ventures. At least part of the time Solomon's kingdom extended from Tiphsah on the Euphrates River in

▲ *et-Tell: Iron Age gate at et-Tell northeast of Sea of Galilee. This is probably Geshur, where Absalom was from. Note the cult center on the right side of the entrance (stairs, basin, and standing stone).*

REBELLIONS AGAINST DAVID: ABSALOM AND SHEBA

→ Absalom's route
→ David's flight from Absalom
■ Cities aiding David in the field
⇢ David's warriors pursue Sheba

0   10 km.
0   10 miles

LIMITS OF ISRAEL PROPER IN WHICH JOAB CONDUCTED A CENSUS FOR DAVID

Mediterranean Sea

Ijon
Tyre
Dan
MAACAH
Hazor
GESHUR
Geshur
Acco
Sea of Galilee
Yarmuk R.
Megiddo
Gilead
Sharon Plain
Beth Shan
Ramoth Gilead
Jordan R.
Shechem
Jabbok R.
Joppa
Aphek
Jazer
Rabbah of the Ammonites
Gezer
Jerusalem
GAD
PHILISTINES
Shephelah
Gaza
Hebron
Dead Sea
Aroer
Arnon Gorge
Negev of Judah
Beersheba

→ Direction in which the census was taken

0   10 km.
0   10 miles

▲ *Gezer: six-chamber gate at Gezer — probably associated with the Solomonic building program (1 Kings 9:15). View from inside the city looking out — note the drain in the center of the gate over which the road ran.*
▼ *Hazor: double (casemate) wall attached to the six-chambered (Solomonic) gate*

the northeast to the border of Egypt in the southwest (1 Kings 4:21 – 24; 2 Chron 9:26; map p. 74). Solomon also sought to solidify his control of the trade routes that passed through Israel. Along the major International Highway he fortified the strategic cities of Hazor, Megiddo, and Gezer (1 Kings 9:15). Gezer was especially important, for it guarded the road that led up to Jerusalem from the west. He also constructed a string of over forty forts, scattered throughout the Negev.

Early in his reign, Solomon controlled the Transjordanian states of Ammon, Moab, and Edom, for his marriages to women of these countries sealed alliances made with their leading families (1 Kings 11:1; cf. 3:1). Solomon took charge

of Gilead and even Damascus for a time. Thus, he controlled all the major routes passing through the southern Levant; by providing overland caravans with food, water, and protection, and by collecting tolls from them, he became very wealthy. Solomon received revenues from merchants and traders as well as from the kings of Arabia and the governors of the land (2 Chron 9:14). He also received a royal visit from the Queen of Sheba, who came bearing gifts.

Almost half of Solomon's rule was taken up with his two great building projects in Jerusalem. The temple, begun in the fourth year of his reign (966 BC), took seven years to build, while his palace took thirteen years. The limestone used in

Mediterranean Sea

Damascus

Tyre

Dan

Hazor

Geshur

Acco

Cabul

Sea of Galilee

Megiddo

Beth Shan

Gilead

Ramoth Gilead

Shechem

Succoth

Cedar wood

Zarethan

Jordan R.

AMMON

Tel Qasile

Lower Beth Horon

Upper Beth Horon

Rabbah of the Ammonites

Joppa

Gezer

Gibeon

Heshbon

Beth Shemesh

Jerusalem

Baalath

Dead Sea

Gaza

Hebron

Aroer

PHILISTINES

Arad

MOAB

Beersheba

Kir Moab

Fortresses in the Negev

Tamar

Bozrah

Kadesh Barnea

Kh. en-Nahas

EDOM

Timnah Copper Mines

Ezion Geber

Solomon's Kingdom

Shipping route

Fortified by Solomon

0   20 km.

0   20 miles

SOLOMON'S KINGDOM AND MAJOR TRADE ROUTES

KUE

Carchemish
Haran
Nineveh
Ain Dara
Tiphsah

ELISHAH
Hamath
Arvad
Tadmor
Euphrates R.
Tigris R.

Byblos
Lebo Hamath

Great Sea
Damascus
Tyre
ARAM
Hazor
Babylon

Megiddo
Joppa
Gezer
Gaza
Jerusalem

Dumah

Memphis

EGYPT
Ezion
Geber

Tema

Nile R.

ARABIA

Red Sea

To Sheba

Solomon's sphere of influence

Major trade routes

Water trade routes

0   100 km.
0   100 miles

the two buildings was cut from the Judean hills. The timber — choice cedar and pine logs — was supplied by Hiram of Tyre. After being cut in the mountains of Lebanon, the logs were floated down the Mediterranean to Joppa (1 Kings 5:1 – 12; 2 Chron 2:3 – 16); from there they were transported up to Jerusalem, probably via the Gezer – Beth Horon road. The Israelites remembered the Solomonic age as a time when luxury items of gold and silver and cedar abounded in the capital (1 Kings 10:14 – 27; 2 Chron 1:15; 9:13 – 24).

In spite of all that wealth, however, it seems that Solomon had to resort to a program of taxing many of his subjects to maintain his large court and lavish lifestyle. He divided all of Israel north of Jerusalem into twelve administrative districts. Evidently Judah, the tribe from which David and Solomon came, was free from this burden, and the hard feelings that resulted probably fueled the north's eventual desire for independence. Each district had to provide food for the king and his household for one month every year (1 Kings 4:7, 22 – 23, 27 – 28).

Slowly, the empire began to show signs of weakness. For example, Hiram of Tyre supplied Solomon with lumber and gold (1 Kings 9:11 – 14), workmen (5:18; 7:13 – 14), and ships and sailors for maritime ventures on the Red Sea (9:27 – 28). In return, Solomon provided Hiram with agricultural products (5:11). But Solomon also gave Hiram twenty cities from the tribe of Asher in the Plain of Acco area (9:11 – 14), most likely because of an unpaid debt.

Toward the end of his reign, some of the leaders of the northern tribes became dissatisfied with him, as seen in Jeroboam's rebellion against Solomon (1 Kings 11:26 – 40). Externally, Hadad the Edomite became Solomon's adversary to the southeast (vv. 14 – 22), while to the northeast Rezon of Damascus became a leader of rebels in Damascus (vv. 23 – 25). Solomon also began to worship some of the foreign deities he had introduced into Jerusalem in deference to his foreign wives (vv. 1 – 8). It is no wonder that the kingdom collapsed almost immediately after Solomon's death.

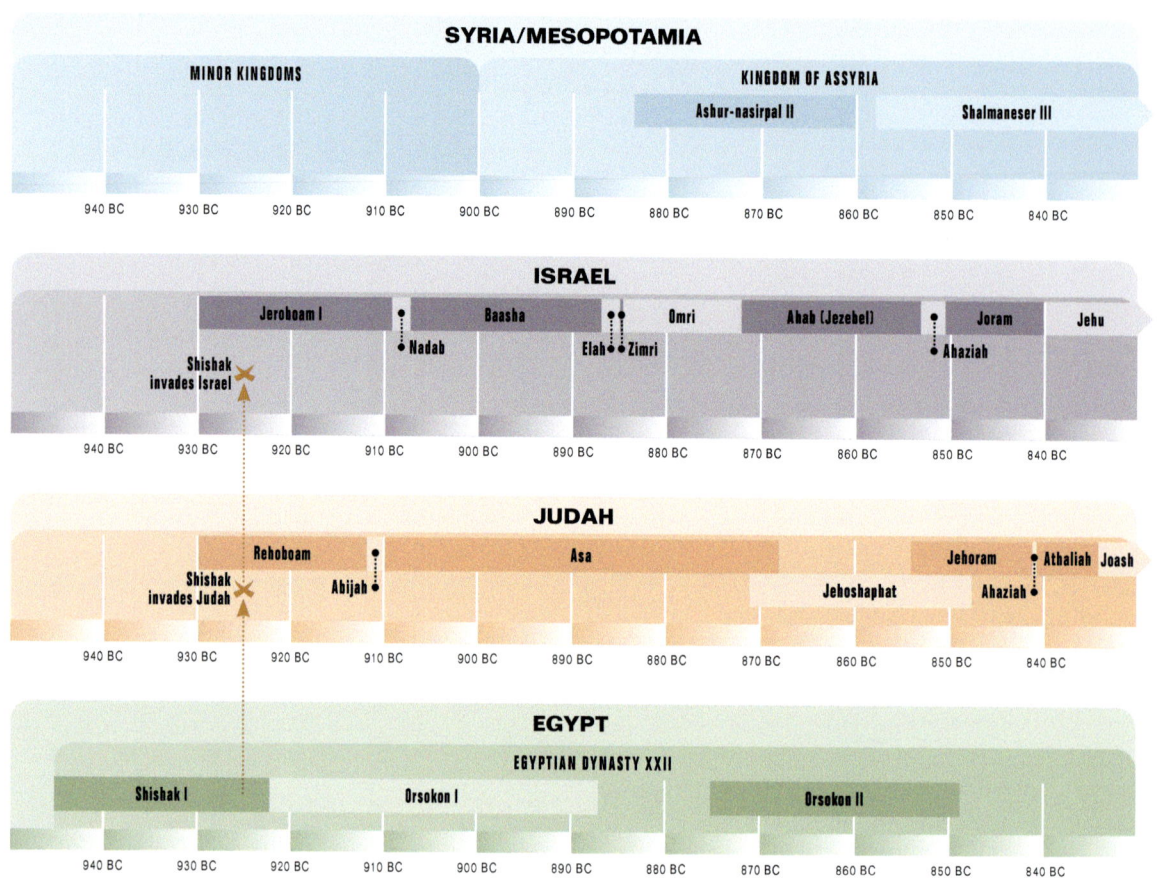

SYRIA/MESOPOTAMIA

MINOR KINGDOMS		KINGDOM OF ASSYRIA		
		Ashur-nasirpal II	Shalmaneser III	

940 BC · 930 BC · 920 BC · 910 BC · 900 BC · 890 BC · 880 BC · 870 BC · 860 BC · 850 BC · 840 BC

ISRAEL

Jeroboam I · Nadab · Baasha · Elah · Zimri · Omri · Ahab (Jezebel) · Joram · Ahaziah · Jehu

Shishak invades Israel

940 BC · 930 BC · 920 BC · 910 BC · 900 BC · 890 BC · 880 BC · 870 BC · 860 BC · 850 BC · 840 BC

JUDAH

Rehoboam · Abijah · Asa · Jehoshaphat · Jehoram · Ahaziah · Athaliah · Joash

Shishak invades Judah

940 BC · 930 BC · 920 BC · 910 BC · 900 BC · 890 BC · 880 BC · 870 BC · 860 BC · 850 BC · 840 BC

EGYPT

EGYPTIAN DYNASTY XXII

Shishak I · Orsokon I · Orsokon II

940 BC · 930 BC · 920 BC · 910 BC · 900 BC · 890 BC · 880 BC · 870 BC · 860 BC · 850 BC · 840 BC

THE DIVIDED KINGDOM

After Solomon's death, his son and successor, Rehoboam, traveled north to the Israelite tribal center at Shechem in order to secure the continued allegiance of the northern tribes (1 Kings 12:1 – 19; 2 Chron 10:1 – 19). But when the latter demanded relief from their oppressive tax burden, Rehoboam's response was that he would increase this burden. At this point the northern tribes rejected the Davidic dynasty and appointed Jeroboam as their first king. Rehoboam fled for his life back to Jerusalem; thus began the period of the Divided Kingdom (930 – 722 BC).

Jeroboam established worship centers at Dan and Bethel (1 Kings 12:26 – 33), located at the extremities of his country. At Bethel he sought to "detour" any worshipers heading to the temple in Jerusalem. At this time many Levites left their allotted cities in northern Israel and moved south to Judah (2 Chron 11:13 – 14).

The Egyptian pharaoh Shishak, seeing the weakness of the divided kingdom, made plans to invade both Judah and Israel. In response, Rehoboam constructed fifteen fortresses in Judah (2 Chron 11:5 – 12) to ward off the attack. These forts were located in the Shephelah, the southern hill country, and along the edge of the Judean Desert; thus, within five years of Solomon's death, the "empire" of his son was confined to the Hill Country of Judah!

Shishak's invasion of Judah is briefly described in the Bible (1 Kings 14:25 – 31; 2 Chron 12:1 – 11), but his own

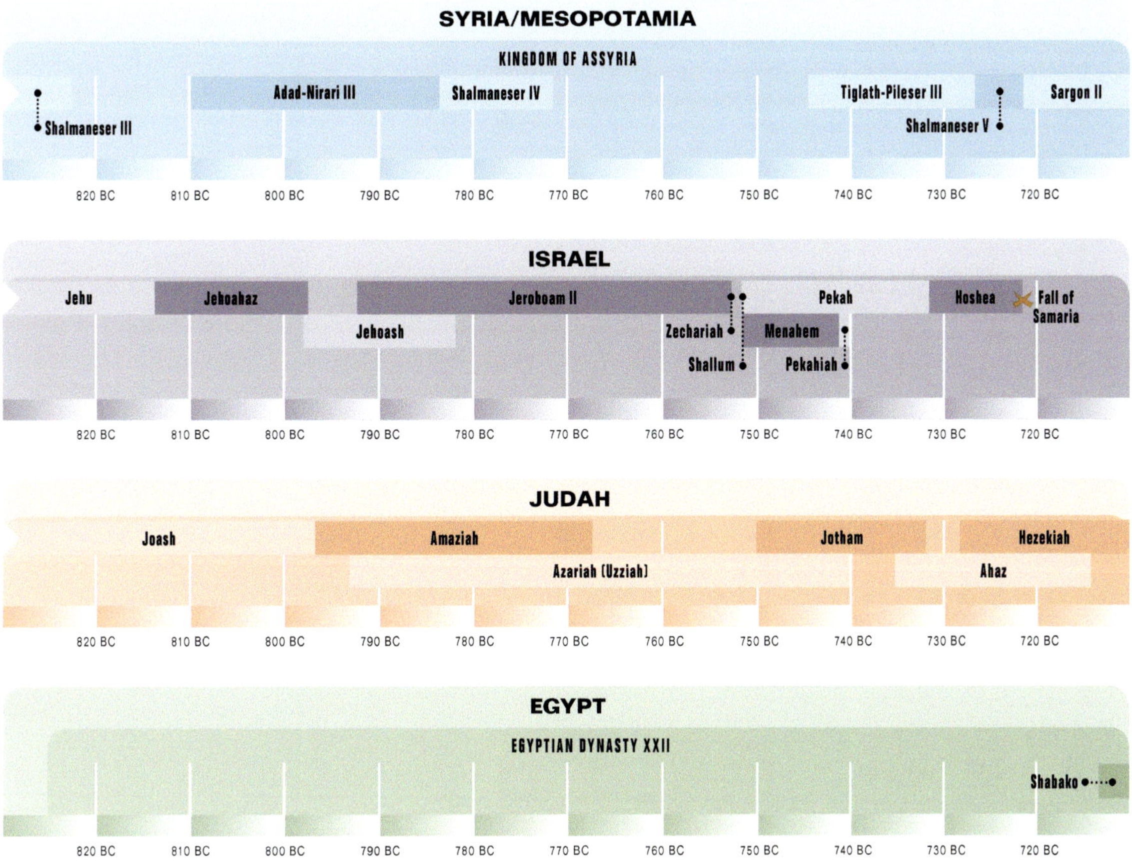

SYRIA/MESOPOTAMIA

KINGDOM OF ASSYRIA

Shalmaneser III • • Adad-Nirari III • Shalmaneser IV • Tiglath-Pileser III • Shalmaneser V • Sargon II

| 820 BC | 810 BC | 800 BC | 790 BC | 780 BC | 770 BC | 760 BC | 750 BC | 740 BC | 730 BC | 720 BC |

ISRAEL

Jehu • Jehoahaz • Jehoash • Jeroboam II • Zechariah • Shallum • Menahem • Pekahiah • Pekah • Hoshea • Fall of Samaria

| 820 BC | 810 BC | 800 BC | 790 BC | 780 BC | 770 BC | 760 BC | 750 BC | 740 BC | 730 BC | 720 BC |

JUDAH

Joash • Amaziah • Azariah (Uzziah) • Jotham • Ahaz • Hezekiah

| 820 BC | 810 BC | 800 BC | 790 BC | 780 BC | 770 BC | 760 BC | 750 BC | 740 BC | 730 BC | 720 BC |

EGYPT

EGYPTIAN DYNASTY XXII

Shabako • • • •

| 820 BC | 810 BC | 800 BC | 790 BC | 780 BC | 770 BC | 760 BC | 750 BC | 740 BC | 730 BC | 720 BC |

▼ *Dan: podium (partially reconstructed) where the king sat on a throne at the entrance to the city gate*

▼ *Dan: high place for worship. Jeroboam originally put "golden calves" at both Bethel and Dan (1 Kings 12:28 – 30).*

Map labels:

Mediterranean Sea

Sidon
Damascus
PHOENICIA
Tyre
Dan
ARAM DAMASCUS
Hazor
Acco
Geshur
GESHUR
Sea of Galilee
Karnaim
Dor
Shunem
Beth Shan
Megiddo
Taanach
Ramoth Gilead
Rehob
Samaria
ISRAEL
Tirzah
Shechem
Penuel
Rabbah of the Ammonites
Jordan R.
Bethel
AMMON
Gezer
Gibeon
Ashdod Yam
Ekron
Jerusalem
Ashdod
Gath
JUDAH
Ashkelon
PHILISTIA
Gaza
Hebron
Dead Sea
Arnon R.
Sharuhen
Raphia
Arad
MOAB
Beersheba
Zered R.
Tamar
Sela
Fortresses in Negev
Bozrah
Kadesh Barnea
E D O M
Teman
Rekem
Ezion Geber
Elath

Scale: 0 — 20 km. / 0 — 20 miles

Legend:
- Kingdom of Israel
- Kingdom of Judah
- ◆ Jeroboam's worship centers
- → Shishak's invasion

inscription documents it more fully. The invasion took place in Rehoboam's fifth year (925 BC), and in one section of his inscription he describes the conquest in the hill country from Jerusalem to the Jezreel Valley. The second section of Shishak's inscription describes his conquest of some eighty-five settlements in the Negev. Thus, control of the major trade routes through Israel slipped from Israelite and Judean hands to the Egyptians.

It took about twenty years for Israel and Judah to resolve the dispute over their border. Rehoboam's son Abijah invaded Israel and captured Bethel, Jeshanah, and Ephron (2 Chron 13:2 – 20). The northern king Baasha responded by pushing the border south to Ramah. In turn, the Judean king Asa felt he needed assistance in meeting this Israelite threat and appealed to Ben-Hadad, king of Aram Damascus, who opened a northern front against Israel (1 Kings 15:16 – 22; 2 Chron 16:1 – 6). Thus, Baasha had to abandon his southward expansion plans, and Asa pushed their common border northward. The resulting boundary, which left Bethel in Israel and Mizpah and Geba in Judah, formed the traditional boundary between the northern and southern kingdoms.

The northern kingdom was characterized by instability. Its nineteen kings came from nine different families, and eight of its kings were either assassinated or committed suicide. This instability is also illustrated by the fact that Israel had four capitals (map p. 80). Shechem was initially selected because of its long history as a tribal and religious center. From there the capital moved to Penuel, and then to Tirzah. Later, King Omri purchased the hill of Shemer and built a new capital there — Samaria (1 Kings 16:23 – 24).

Samaria was situated closer to the coastal plain and open to outside influences. Indeed, Israel's treaty with the king of the Sidonians indicates its outward-looking orientation. The marriage of Ahab and Jezebel (a Sidonian) cemented this alliance. Ahab's building of a temple and altar in Samaria for Jezebel's god Baal indicates that external influences on the northern kingdom were religious as well (1 Kings 16:31 – 33).

But Israel's relations with the Arameans of Damascus deteriorated. One of the reasons for this was that Israel controlled cities such as Dan, Ijon, Abel Beth Maacah, and Hazor that sat on the east – west caravan route from Damascus to the Mediterranean (map p. 80). In addition, Israel controlled the lucrative north – south incense route along the King's Highway in Transjordan and the route that lead from Ramoth Gilead to the port of Acco. In effect, Israel collected caravan transit revenues rather than Damascus. There were at least thirteen battles between Israel and the Arameans — several occurring near the strategic city of Ramoth Gilead.

On a number of occasions the Arameans pressed their advantage to the gates of Samaria, but each time they were driven off to lands east of the Jordan River. But it eventually became necessary for Israel and Aram to join together along with other nations of the Levant to meet a growing Assyrian threat led by Shalmaneser III. This coalition met him in battle at Qarqar on the Orontes in 853 BC, where they temporarily blunted the Assyrian threat (map p. 82).

Mark Connally

▲ *Karnak, Egypt: Shishak conquering his foes (lower right) who are pleading for mercy. In the lower left are cartouches of captured enemies/cities (1 Kings 14:25 – 28).*

But soon after the battle of Qarqar, Ben-Hadad and Ahab were fighting each other again; Ahab was killed in the battle at Ramoth Gilead (1 Kings 22:29 – 37; 2 Chron 18:28 – 34). With his death, Moab, Israel's vassal, gained its independence at the expense of the Israelite tribes of Gad and Reuben (2 Kings 3:4 – 27),

Ahab's son Joram bore the brunt of Aram's continuing attacks (2 Kings 6:24 – 7:8). While Joram was recuperating

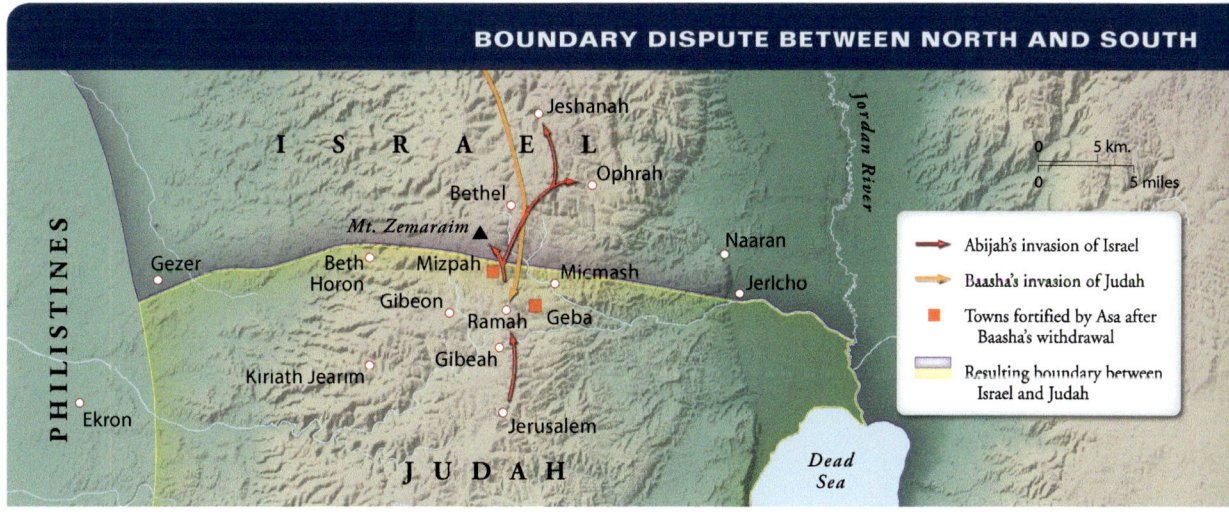

BOUNDARY DISPUTE BETWEEN NORTH AND SOUTH

I S R A E L

Jeshanah

Ophrah

Bethel

Jordan River

Mt. Zemaraim ▲

Naaran

Gezer

Beth Horon

Mizpah

Micmash

Jericho

Gibeon

Ramah   Geba

PHILISTINES

Kiriath Jearim

Gibeah

Ekron

Jerusalem

Dead Sea

J U D A H

0   5 km.
0   5 miles

→ Abijah's invasion of Israel

→ Baasha's invasion of Judah

■ Towns fortified by Asa after Baasha's withdrawal

▭ Resulting boundary between Israel and Judah

THE NORTHERN KINGDOM—ISRAEL (930–722 BC)

Worship centers built by Jeroboam
Royal centers in Israel
Important Mediterranean shipping ports
International routes
Local or regional routes

PHOENICIANS

ARAMEANS

Sidon

Damascus

Tyre

Ijon

Dan

Abel Beth
Maacah

Kedesh

Iron

Hazor

Acco

Kinnereth

Karnaim

Sea of
Galilee

Ashtaroth

Hannathon

Aphek

Yarmuk R.

Edrei

Dor

Megiddo

Jezreel

Lo Debar

Aruna

Taanach

Beth Shan

Ramoth Gilead

Ibleam

Rehob

Jabesh Gilead

Frequent battleground
between the Arameans
of Damascus and Israel
for control of trade route

Socoh

Tirzah

Zaphon

Jordan R.

Great Sea
(Mediterranean Sea)

Samaria

Succoth

Penuel

Mahanaim

Jabbok R.

Shechem

Adam

ISRAEL

Joppa

Aphek

Rabbah of the
Ammonites

AMMONITES

Beth
Horon

Bethel

Zemaraim

Gibbethon

Gezer

Gibeon

Ashdod

Ekron

Kiriath
Jearim

Jerusalem

Ashkelon

Gath

PHILISTINES

JUDAH

Gaza

Hebron

En Gedi

Sea of
Arabah
(Salt Sea)

Dibon

Moabite stone discovered here.
Mesha, King of Moab, in
conflict with Kings of Israel

Aroer

Arnon Gorge

0   10 km.
0   10 miles

MOABITES

from his wounds in Jezreel, Jehu staged a coup, executing the king, Jezebel, and all the descendants of Omri/Ahab (841 BC; 2 Kings 9 – 10) as well as the priests and worshipers of Baal (2 Kings 10). But it is known from Assyrian records (*ANET*, 280 – 81) that Jehu paid tribute to Shalmaneser III. Near the end of Jehu's reign, Hazael the Aramean invaded Israel, capturing Transjordanian lands (2 Kings 10:32 – 33).

Judah exhibited much more stability during its 345-year history. Judah had nineteen kings, all from the Davidic dynasty. In addition, Jerusalem had been firmly established as the religious and political capital of the southern kingdom.

The Chronicler's account of the reigns of successive Judean kings illustrates the theological principle that their expressions of fidelity and trust toward God led to blessing, prosperity, and strength, while their disobedience usually led to disaster, destruction, defeat, and eventually deportation. For example, early in Asa's reign, expressions of his trust in God were followed by victory over invading hordes (2 Chron 14). But later, Asa's lack of trust in the power of God to deal with Baasha's aggressive move southward and his alliance with Ben-Hadad of Damascus led to his ignominious death (16:7 – 14).

Asa's successor, Jehoshaphat, instituted religious and legal reforms. In addition, garrison cities, forts, and store cities were constructed and manned (2 Chron 17). Jehoshaphat's power was such that the Philistines to the west and the Arabians to the south and east brought him tribute — possibly because he was again exercising Judean control over portions of the international trade routes. Later, the Moabites, Ammonites, and Meunites invaded Judah from the east but were defeated by divine help because of Jehoshaphat's piety (2 Chron 20).

Nevertheless, Jehoshaphat established close relations with Israel. In fact, his son Jehoram married Athaliah, a daughter of Ahab and Jezebel (2 Kings 8:18; 2 Chron 21:6),

At various times in its history, Judah was oppressed by its neighbors to the west, south and east. Conversely, when Judah was strong, it expanded at their expense.

853 BC Battle between Shalmaneser III and 12 Levantine Kings, including Ahab

The expansion of the Assyrian Kingdom:

	Under Shalmaneser III
	Under Tiglath-pileser III
	Under Esarhaddon
	Under Asshurbanipal

0   100 km.

0   100 miles

and Jehoshaphat even fought alongside Ahab in the battle of Ramoth Gilead. Indeed, every recorded joint venture between Jehoshaphat and Israel was unsuccessful.

After the death of Jehoram, a wicked king who experienced a number of reversals, the royal Davidic line was almost annihilated by the wicked queen mother, Athaliah (2 Kings 11). But one child from the Davidic line was rescued. Early in his rule, Joash began a number of religious reforms (2 Kings 12:1 – 12; 2 Chron 24:1 – 16). Later, however, he forsook God for the worship of the Asherim and idols (2 Chron 24:17 – 22) and suffered military reverses.

Disgruntled elements assassinated Joash (2 Kings 12:19 – 21; 2 Chron 24:25 – 27), and his son Amaziah ruled in his place. Amaziah managed to subdue Edom (2 Kings 14:7; 2 Chron 25:1 – 15). However, Israel invaded Judah and destroyed portions of the walls of Jerusalem (2 Kings 14:8 – 14;

ASSYRIAN PROVINCES AFTER THE FALL OF THE NORTHERN KINGDOM

△ Cities captured by Tiglath-pileser III

△ Cities captured by Shalmaneser V and Sargon II

▢ Assyrian provinces in the days of Tiglath-pileser III

▢ Assyrian provinces added in the days of Shalmaneser V and Sargon II

Mediterranean Sea

MASUATE

SUBITE

Sidon

Damascus

Ijon

DAMASCUS

Tyre

Janoah

Abel Beth Maacah

Kedesh

Iron

Merom

Janoah?

Hazor

KARNAIM

Acco

Jotbathah

Kanah

Karnaim

Hannathon

Arumah

Sea of Galilee

Ashtaroth

Dor

Megiddo

Beth Shan

Ramoth Gilead

GILEAD

HAURAN

DOR

SAMARIA

Samaria

Succoth

Jordan R.

Aphek

AMMON

Joppa

Bethel

Jericho

Rabbah of the Ammonites

Gibbethon

Gezer

Jerusalem

Heshbon

Bezer

Ashdod Yam

Ekron

Bethlehem

Medeba

Ashdod

Gath

Azekah

Beth Diblathaim

Ashkelon

Lachish

Dibon

JUDAH

Hebron

En Gedi

MOAB

Aroer

Gaza

ASHDOD

Gerar

Beersheba

Dead Sea

Kir Moab

Raphia

Arad

Aroer

Zoar

EDOM

Bozrah

0   10 km.

0   10 miles

2 Chron 25:17 – 24). Amaziah's political support dwindled, and he was assassinated (2 Kings 14:18 – 20; 2 Chron 25:26 – 28).

Amaziah was succeeded by his son Azariah (also called Uzziah). In the west this "pious king" captured Gath, Jabneh, and Ashdod (2 Chron 26:6 – 8). To the south and southwest the Arabs in Gur Baal and the Meunim paid tribute to him, as did the Ammonites on the east. In the south he constructed forts in the wilderness, subdued the Edomites, and rebuilt Elath on the Red Sea (2 Kings 14:22; 2 Chron 26:2).

Jotham, Azariah's son, continued in the ways of his father. But toward the end of his reign, Pekah of Israel and Rezin of Aram Damascus attacked Judah. The Judeans appealed to the Assyrian monarch Tiglath-Pileser III for assistance, and he responded by attacking Israel (2 Kings 16:7 – 10; 15:29; 2 Chron 28:16, 20 – 21. Although the Judeans were relieved of immediate Israelite/Aramean pressures, they later suffered at the hands of their Assyrian ally.

During Ahaz's reign religious and political conditions in Judah deteriorated (2 Kings 16:2 – 4; 2 Chron 28:1 – 4). These religious setbacks were paralleled by military reversals: an Israelite/Aramean invasion (2 Chron 28:6), an Edomite revolt (2 Kings 16:5 – 6), and a Philistine invasion of the Shephelah and the Negev (2 Chron 28:17 – 19; map p. 81).

From roughly 800 BC until 740 BC the Assyrians were pre-occupied elsewhere in the Near East, and the Israelite kingdom was able to flourish. The peak of Israelite expansion and prosperity was reached during the long rule of Jeroboam II (793 – 753 BC). The area of his influence stretched from Lebo Hamath in the north to Judah and the Sea of the Arabah (= Salt Sea) in the south (2 Kings 14:25 – 29). Indeed, when the territories controlled by the Judean king Azariah/Uzziah and those controlled by Jeroboam II are considered together, their combined area almost reached Solomonic proportions.

Jeroboam II's two successors ruled for a combined total of seven months; both were assassinated. During the reign of Menahem (752 – 742 BC) the Assyrian menace was again felt. Menahem's successor, Pekah, bore the brunt of the initial Assyrian onslaught.

Biblical and especially Assyrian records describe Tiglath-Pileser III's invasions (2 Kings 15:29 – 30; ANET, 282 – 84). In 734 BC he marched down the Mediterranean coast all the way to Gaza and the Brook of Egypt (maps pp. 82, 83). In 733 BC his army returned, capturing northern Israelite cities and Gilead. Finally in 732 BC, he attacked and captured Damascus. Rezin was deposed and Pekah assassinated; in the latter's stead, Tiglath-Pileser III put Hoshea (732 – 722 BC) on the Israelite throne as a puppet king.

Shortly after Tiglath-Pileser III died (727 BC), Hoshea revolted against the Assyrians, who responded by laying siege to Samaria under Shalmaneser V. Samaria eventually fell to the Assyrians (2 Kings 17:4 – 6; 18:9 – 11). The fall of Samaria was a traumatic experience for the Israelites; Sargon II of Assyria began a series of deportations (2 Kings 17:6; 18:10 – 11). In their place, the Assyrian rulers settled foreign peoples (17:24). These newcomers brought with them their worship of pagan deities; yet they also attempted to worship the "god of the land" (i.e., Yahweh; vv. 25 – 41). Evidently these newcomers, with their syncretistic blend of religions, were the forerunners of the religious/ethnic group later known as the Samaritans.

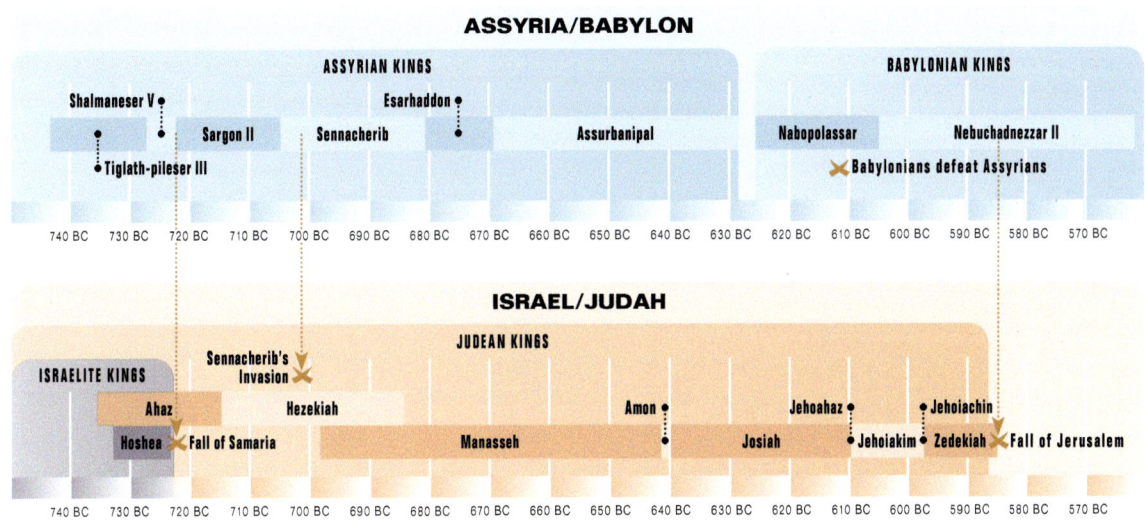

ASSYRIA/BABYLON

ASSYRIAN KINGS — Shalmaneser V, Tiglath-pileser III, Sargon II, Sennacherib, Esarhaddon, Assurbanipal

BABYLONIAN KINGS — Nabopolassar, Nebuchadnezzar II, ✕ Babylonians defeat Assyrians

740 BC · 730 BC · 720 BC · 710 BC · 700 BC · 690 BC · 680 BC · 670 BC · 660 BC · 650 BC · 640 BC · 630 BC · 620 BC · 610 BC · 600 BC · 590 BC · 580 BC · 570 BC

ISRAEL/JUDAH

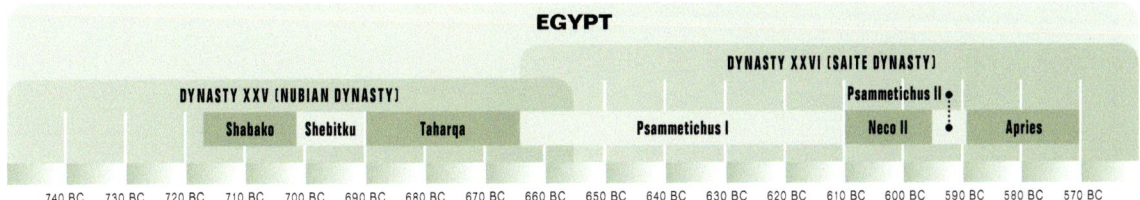

JUDEAN KINGS

ISRAELITE KINGS — Ahaz, Hoshea, ✕ Fall of Samaria, Sennacherib's Invasion ✕, Hezekiah, Manasseh, Amon, Josiah, Jehoahaz, Jehoiakim, Jehoiachin, Zedekiah, ✕ Fall of Jerusalem

740 BC · 730 BC · 720 BC · 710 BC · 700 BC · 690 BC · 680 BC · 670 BC · 660 BC · 650 BC · 640 BC · 630 BC · 620 BC · 610 BC · 600 BC · 590 BC · 580 BC · 570 BC

EGYPT

DYNASTY XXV (NUBIAN DYNASTY) — Shabako, Shebitku, Taharqa

DYNASTY XXVI (SAITE DYNASTY) — Psammetichus I, Neco II, Psammetichus II, Apries

740 BC · 730 BC · 720 BC · 710 BC · 700 BC · 690 BC · 680 BC · 670 BC · 660 BC · 650 BC · 640 BC · 630 BC · 620 BC · 610 BC · 600 BC · 590 BC · 580 BC · 570 BC

JUDAH ALONE

▼ *Jerusalem: eighth-century wall built to defend the western hill of Jerusalem (2 Kings 22:14; Isa 22:10). Twenty feet thick!*

The Assyrian victory over Damascus and the northern kingdom of Israel brought the Judean kingdom temporary relief from military pressures, but Ahaz's promotion of pagan religious practices almost ensured that the southern kingdom would also fall under God's judgment. The Edomites invaded Judah while the Philistines captured cities in or near the Shephelah (2 Chron 28:17 – 19). At this time many Israelites from the northern kingdom moved south to avoid the onslaught of the Assyrians. Jerusalem grew from a city of 37 acres to 150 acres and many new settlements sprang up in the Hill Country of Judah.

With the death of Ahaz in 715 BC, Hezekiah's sole rule began. Very early he initiated a series of religious reforms. High places were torn down, sacred stones smashed, Asherah poles destroyed,

Legend:
- Sennacherib's invasion of Judah
- Assyrian advance from the north (Isaiah 10:28–32)
- Tirhakah, king of Kush
- See city list of Judah (Joshua 15:21–63)

and even the bronze snake that Moses had set up in the wilderness (Num 21:5–9) was smashed (2 Kings 18:3–7; 2 Chron 29:2–19). In addition, the temple was reconsecrated and a great Passover was celebrated in Jerusalem (2 Chron 29:20–30:27).

The death of the Assyrian king Sargon II in 705 BC was the signal for many countries in the Near East to attempt to assert their independence. Hezekiah, for example, attacked the Philistines and regained territory as far as Gaza (2 Kings 18:8). In the process he deposed petty kings still loyal to Assyria and replaced them with rulers more to his liking.

Hezekiah must have anticipated that sooner or later the new Assyrian king, Sennacherib, would respond. Hezekiah stationed garrisons throughout the Hill Country of Judah and in the Shephelah. They were provided with officers, weapons, shields, and food supplies. Hezekiah's preparations in and around Jerusalem were especially noteworthy; he dug a 1,750-foot-long water tunnel through rock and built a massive city wall on the newly settled western hill.

Sennacherib's response in 701 BC is one of the best-documented events in the ancient world. Scripture describes his invasion from the Judean standpoint (2 Kings 18 – 20; 2 Chron 32:1 – 23; Isa 36 – 39). From the Assyrian standpoint the Prism of Sennacherib (*ANET*, 287 – 88) describes his invasion in great detail. In addition, stone reliefs that lined the throne room of his palace in Nineveh depict various facets of his Judean campaign, including his siege of Lachish.

After marching westward from Assyria to the Mediterranean Sea, Sennacherib proceeded south, capturing cities along the Phoenician coast and in northern Philistia. The Egyptians and Ethiopians, who had responded to Hezekiah's call for help, were defeated in the Plain of Eltekeh. Sennacherib proceeded farther south and captured the Judean stronghold of Lachish. Excavations at Lachish have found from 6 to 10 feet of ash and debris from this destruction. From his camps at Lachish and Libnah, Sennacherib sent his representatives to demand the submission of Jerusalem (2 Kings 18:17).

In the end, however, Judah's deliverance came when God sent his angel to destroy 185,000 Assyrians (19:35 – 36; 2 Chron 32:21 – 22; Isa 37:36). Obviously, Sennacherib did not mention this disastrous loss of troops in his inscriptions. Instead, he emphasized how he besieged and conquered forty-six strong cities and countless villages, how he confined Hezekiah to Jerusalem "like a bird in a cage," and how many kings in the area, Hezekiah included, sent him tribute.

In spite of Sennacherib's losses on his Judean campaign, Assyria remained powerful. In the 670s BC the Assyrians were again active along the Mediterranean coastline, attempting to invade Egypt. In the biblical text Manasseh, Hezekiah's successor, is primarily noted for the pagan religious rites that he reinstituted in Judah — including high places, altars for Baals, making an Asherah pole, worshiping the hosts of heaven, making children pass through the fire, and placing idols in the temple (2 Kings 21:2 – 9, 16; 2 Chron 33:1 – 20).

Amon, the wicked son of Manasseh, ruled for only two years. He was followed by Josiah, who ruled for thirty-one years. In his eighth year (632 BC) Josiah sought God, and in his twelfth year (628 BC) he began to purge Judah and Jerusalem of their high places, Asherahs, and images. Josiah exerted some influence over some of the Israelites who remained in what used to be the northern kingdom. The heartland of the major area under his control, however, was probably confined to the Hill Country of Benjamin and Judah (2 Kings 23:8).

London, British Museum

▲ *Assyrians assaulting the Judean city of Lachish during the days of Hezekiah. Note the tower of the city, the battering ram, archers, torches, and lower right, Judeans being impaled on poles. Relief from Sennacherib's palace at Nineveh.*

▼ *Jerusalem: several "false starts" near the center of Hezekiah's Tunnel, where two work gangs met.*

London, British Museum

▲ *Sennacherib on his throne receiving booty and prisoners from his Judean campaign*

During the final years of Manasseh's reign the Babylonians initiated the process that led to the downfall of Assyria. By 620 BC Nabopolassar had established his control of southern Mesopotamia and was ready to move against Assyria. In addition, in the northeast the powerful Medes were pressuring the Assyrians. In 614 BC the Medes captured and destroyed the Assyrian city of Asshur. The Medes and Babylonians then combined forces and in 612 BC conquered Nineveh. The last Assyrian ruler, Ashur-uballit II, joined by his Egyptian ally, Neco II, tried to regain power. In 609 BC, as Neco was heading

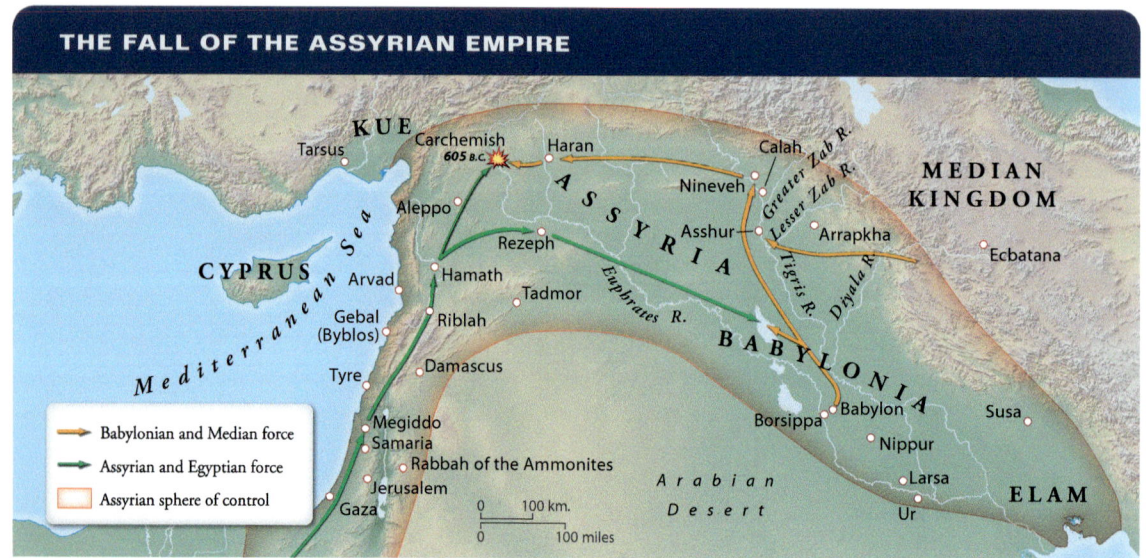

THE FALL OF THE ASSYRIAN EMPIRE

north to assist the Assyrians, Josiah, evidently siding with the Babylonians, tried to impede Neco's march at Megiddo. There the godly Judean king was killed (2 Kings 23:29 – 30; 2 Chron 35:20 – 27).

Judah vacillated between allegiance to Egypt or to Babylon. At Carchemish in 605 BC, the Babylonians defeated the Egyptians, and Judah became a Babylonian vassal. Nebuchadnezzar deported some of the talented upper-class Judeans to Babylon — recorded in Scripture as the first of four deportations of Judeans (Dan 1:1; Jer 46:2; compare 52:28 – 30). This new world empire is called the Neo-Babylonian Empire (605 – 539 BC).

King Jehoiakim again revolted against Babylonian overlordship. Nebuchadnezzar invaded Judah in 597 BC. Just before Nebuchadnezzar's capture of the "city of Judah" (= Jerusalem) on March 16, 597 BC, Jehoiakim died. The next king, Jehoiachin, ruled for only three months (598 – 597 BC) and presided over the fall of the city. In this second deportation (2 Kings 24:13 – 16), he and more than 10,000 Judeans, along with treasures from the temple and palace, were taken captive to Babylon.

Nebuchadnezzar placed Mattaniah, Jehoiachin's uncle, on the throne and changed his name to Zedekiah. Zedekiah ruled for eleven years, at first as a loyal Babylonian vassal. Then, possibly around 588 BC, when the Egyptian king Hophra/Apries (589 – 570 BC) led a military expedition to Tyre and Sidon, Zedekiah joined (Jer 27:1 – 11) in a revolt against Babylon. Nebuchadnezzar again responded quickly. He first attacked the strategically important Shephelah cities of Lachish and Azekah.

In January of 588 BC, Nebuchadnezzar began the siege of Jerusalem. In July 586 BC Jerusalem fell, and between August 14 and 17 the city was razed and the temple burned by Nebuzaradan, commander of the imperial

The map shows place names including Tarsus, Carchemish, Haran, Nineveh, Aleppo, Rezeph, Calah, Asshur, Arrapkha, Ecbatana, Hamath, Tadmor, Sippar, Cuthah, Borsippa, Babylon, Nippur, Susa, Larsa, Ur, and regions labeled CYPRUS, ASSYRIA, MEDIAN EMPIRE, ARAM, NEO-BABYLONIAN EMPIRE, ELAM, EGYPT, JUDAH, MOAB, EDOM, AMMON, Arabian Desert.

| 0 | 100 km. |
| 0 | 100 miles |

→ Exile from Judah

▼ *Iron Age tomb on the grounds of the École Biblique in Jerusalem. Note the burial bench and the people looking into the repository where the bones were collected.*

guard. Zedekiah was captured, his eyes were put out, and he was deported to Babylon (2 Kings 25:1 – 7).

The independent Israelite/Judean state, which had existed for more than four hundred years, had come to an end. Jerusalem was in ruins, the temple of Yahweh had been destroyed, and the heirs of the once-great Davidic dynasty were prisoners in exile. Had God abandoned his people? What had happened to the glorious promises made to Israel's ancestors? It would take a "second exodus," this time from Babylon rather than from Egypt, for God to deliver his people from the catastrophe that had befallen them.

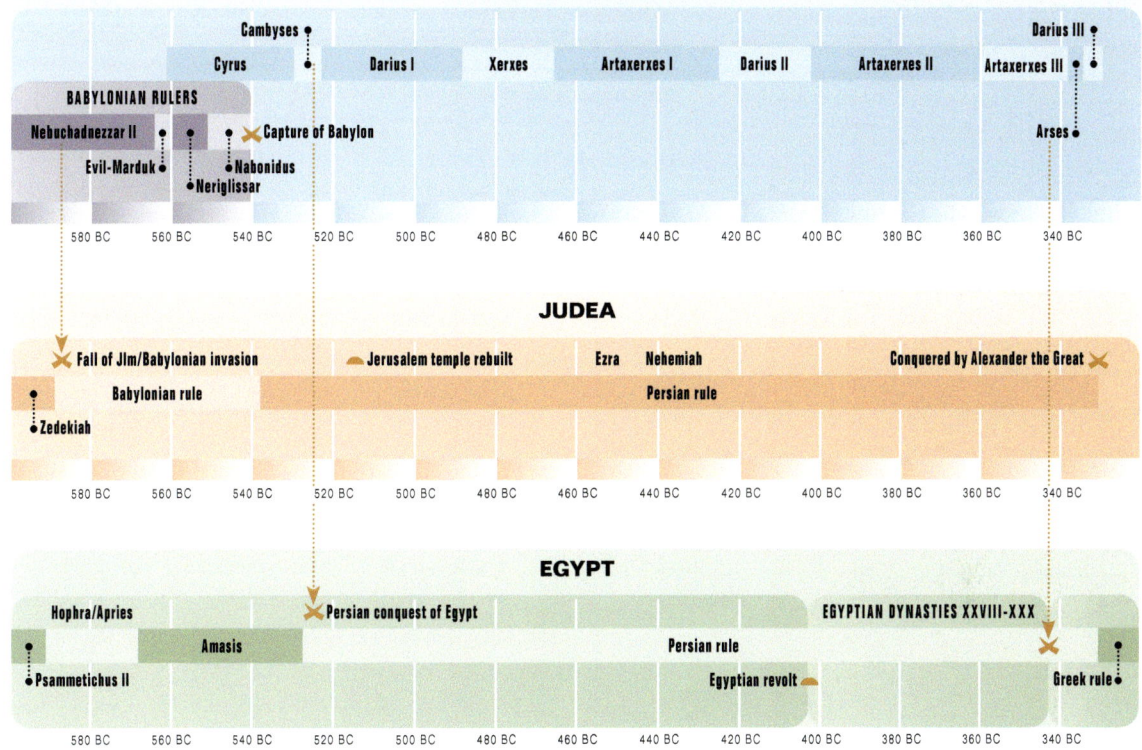

EXILE AND RETURN

With Jerusalem's fall in 586 BC, some Judeans were taken captive to southern Mesopotamia, some remained in the land of Judah, and others fled to neighboring countries — including Ammon, Moab, and Egypt. Gedaliah was appointed governor of those who remained in the land (Jer 40:1 – 41:15). He established his headquarters at Mizpah, north of Jerusalem (map p. 92). Gedaliah's rule ended when Ishmael, sent by the Ammonites (Jer 40:14), murdered Gedaliah.

The Babylonians were still active in the area, and the Judeans were fearful of reprisals (Jer 41:16 – 17). Portions fled to Egypt (map p. 92), taking Jeremiah with them

(42:1 – 44:30). The prophet evidently died in exile in Egypt. In 582 BC the Babylonians took several thousand additional Judeans into exile (52:30).

Specific details about life in Judea from 586 BC to the first return in 538 BC are lacking. Since the Babylonians did not import foreigners to settle areas recently vacated by exiled Judeans, most likely the Edomites moved into the southern portion of the Hill Country of Judah. These people came to be known as the Idumeans.

Life in Babylon for the exiled Jewish populace must have been somewhat depressing, for their religious beliefs were intimately tied up with the land of Israel, from which they were exiled, and with Jerusalem and its temple, which lay in ruins (see Ps 137). In addition, the Davidic dynasty no longer ruled. The question that loomed large in their minds was: Why?

Nevertheless, some portions of the Jewish community seem to have flourished. Indeed, Daniel became an advisor to kings and royalty; Ezekiel moved about; the exiled Judean

JEWISH REFUGEES FLEE TO EGYPT

Map labels:
Great Sea · AMMON · Mizpah · Jeremiah taken to Egypt · JUDAH · Jerusalem · MOAB · Pelusium · Ienysus · Tahpanhes · Migdol · EDOM · Tell el-Maskhuta · Memphis (Noph) · Nile R. · Flight to Egypt · Hermopolis · E G Y P T · Red Sea · Thebes (No-Amon) · Jewish colony established · Elephantine (Yeb) · Aswan (Syene) · 0 100 km. · 0 100 miles

lands. This included the Jews, whom Cyrus permitted to return to Judea and rebuild the temple (2 Chron 36:22 – 23; Ezra 1:1 – 4; 6:3 – 5). A total of 49,697 individuals returned to Judea with Sheshbazzar (Ezra 2:64 – 65).

During this first phase of the first return, around 537 BC, the sacrificial altar was reconstructed, the foundations for the temple were laid, and the Feast of Tabernacles was celebrated (Ezra 1 – 3). But because of the hostility of their enemies (4:1 – 4), work on the temple ceased until the second year of Darius I (ca. 520 BC). The second phase of the return was led by Zerubbabel. Work on the temple was completed by 516 BC (Ezra 4:24 – 6:22).

Internally the Persian Empire was divided into large administrative units called satrapies. Judah was in a satrapy called "Babylon and Beyond the River [Euphrates]." By 520 BC the satrapy had been divided into at least two sub-units, one of which was called "Beyond the River" (NIV "Trans-Euphrates"). It stretched from the Upper Euphrates southwest into northern Sinai.

During the reign of Darius I (522 – 486 BC), the Persian Empire reached its greatest extent, stretching from the Indus River in the east to Thrace and Macedonia in the west. During his reign the minting of coins became common, the legal system of courts and judges was established, a postal system and roads were in full operation, and even a major canal, connecting the Gulf of Suez and the Nile, was completed.

Toward the midpoint of Darius's long reign, trouble began brewing in the Aegean area with the Ionian revolt. Darius suppressed the revolt, and the city of Miletus was destroyed. But in 490 BC a Persian army invading Greece was defeated at Marathon. Other major battles between the Greeks and Persia followed, with Persia finally pulling back (see next chapter).

After the Persian king Xerxes was assassinated, Artaxerxes I (464 – 424 BC) took the throne. He offered additional concessions to the Jews in 458 BC (Ezra 7:7 – 9), which allowed Ezra and an unlimited number of

king Jehoiachin and his entourage received rations from the Babylonians (*ANET*, 308) and was released from prison in 561 BC (2 Kings 25:27 – 30; Jer 52:31 – 34).

The collapse of the Babylonian Empire, like that of the Assyrian one before it, went quickly. King Cyrus moved northwest into Asia Minor, defeating Croesus, the king of Lydia, so that Cyrus's territory stretched from Persia westward to the coast of the Aegean Sea. Cyrus captured Babylon in 539 BC and was hailed as a liberator, treating the populace with leniency. His barrel inscription (*ANET*, 315 – 16) recounts how some peoples exiled by the Babylonians were allowed to return to their home-

London, British Museum

▲ *Clay foundation cylinder (4 in. long), discovered at the ziggurat at Ur in southern Iraq, mentioning both Nabonidus and Belshazzar his son (Daniel 5).*

FALL OF BABYLON; RETURN FROM EXILE

ASIA MINOR

URARTU

Caspian Sea

LYDIA

Aegean Sea

Cyrus to Sardis

Sardis
546 B.C.

CILICIA

Tigris R.

MESOPOTAMIA

Haran

Gozan

Nineveh

Arbela

Ecbatana
550 B.C.

Aleppo

Arpad

Euphrates R.

M E D E S

SYRIA

Hamath
Qatna

Riblah

Tadmor

Opis

Cyrus

Byblos

BABYLONIA

Babylon
539 B.C.

Sidon

Damascus

Nippur

Susa

Mediterranean Sea

Tyre

Acco

AMMON

Samaria

Shechem

Rabbah

ELAM

Ur

Jerusalem

MOAB

Pelusium

EDOM

Nile R.

Memphis

Persian Gulf

E G Y P T

Tema

A R A B I A

Red Sea

DEDAN

Yathrib

Legend

→ Persian conquest of Babylonian Empire

● Concentration of Judean exiles

Return to Zion:

→ Return in the days of Sheshbazzar and Zerubbabel

→ Returns under Ezra and Nehemiah

0   100 km.
0   100 miles

THE PERSIAN EMPIRE

Aral Sea

Jaxartes R.

Danube R.

Black Sea

Sinope

Caspian Sea

SOGDIANA

MACEDONIA

THRACE

Byzantium

Phasis

Oxus R.

HYRCANIA

Trapezus

Marathon

Miletus

CAPPADOCIA

Euphrates R.

Margiana

BACTRIA

Thermopylae

Delphi

LYDIA

ARMENIA

PARTHIA

Tesmes
(Meshed)

GANDHARA

Athens

LYCAURIA

Sardis

Iconium

MEDIA

Taxila

Sparta

IS. OF THE SEA

Derbe

CILICIA

Haran

Arbela

Ecbatana

Damghan

ARIA

CRETE

Xanthos

Tarsus

Asshur

Behistun

SAGARTIA

ARACHOSIA

Gortyna

CYPRUS

Kition

BEYOND THE RIVER

Hamath

Tadmor

Thapsacus

Eshnunna

Der

Susa

Gabae
(Isfahan)

Kerman

INDIA

Indus R.

Tyre

ARABIA

Sippar

SHUSHAN

Pasargadae

Pura

Patala

LIBYA

Ashdod

Damascus

BABYLONIA

Uruk

Persepolis

Sais

Jerusalem

Babylon

Ur

PARSA

Heliopolis

Pelusium

Dumah

MAKA

Mediterranean Sea

EGYPT

Tema

Persian Gulf

Nile R.

Thebes

Dedan

Arabian Sea

Elephantine

Yathrib

Aswan

Red Sea

Legend

● Royal residences

MEDIA Satrapy under Darius I

— Canal built between the Gulf of Suez and the Nile

— Royal Way

0   300 km.
0   300 miles

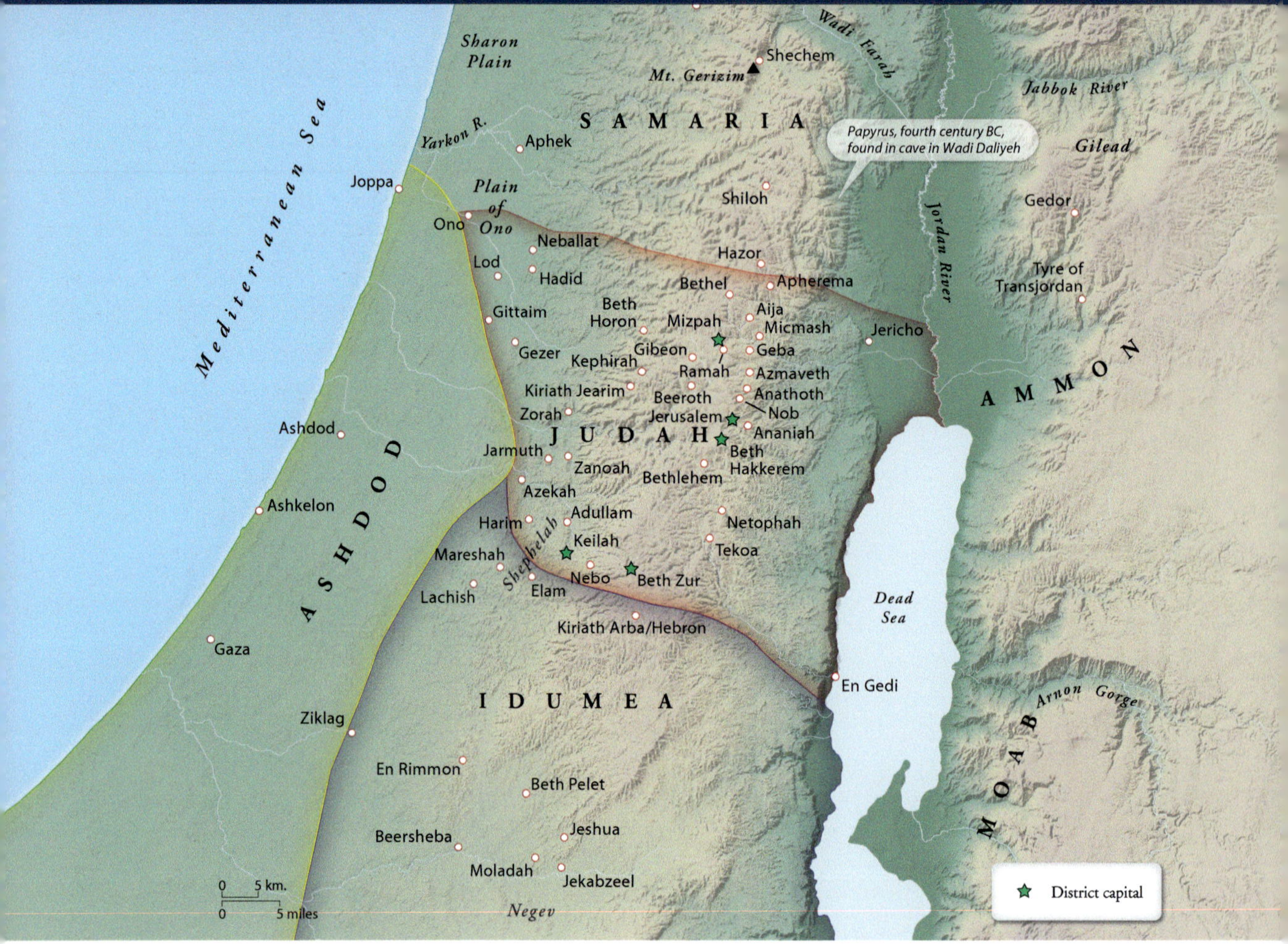

Map: The Province of Judah (Yehud)

Mediterranean Sea

Sharon Plain

Mt. Gerizim · Shechem

Wadi Farah

S A M A R I A

Jabbok River

Gilead

Yarkon R. · Aphek

Papyrus, fourth century BC, found in cave in Wadi Daliyeh

Jordan River

Joppa

Plain of Ono

Gedor

Ono

Neballat

Shiloh

Tyre of Transjordan

Lod

Hadid

Hazor

Jericho

Gittaim

Beth Horon

Bethel

Apherema

Mizpah

Aija

Micmash

Gezer

Kephirah

Gibeon

Geba

A M M O N

Ramah

Azmaveth

Ashdod

Kiriath Jearim

Beeroth

Anathoth

Zorah

J U D A H

Jerusalem

Nob

Jarmuth

Ananiah

Zanoah

Beth Hakkerem

Azekah

Bethlehem

Adullam

Harim

Keilah

Netophah

Mareshah

Shephelah

Tekoa

Lachish

Nebo

Elam

Beth Zur

Dead Sea

Kiriath Arba/Hebron

I D U M E A

En Gedi

Gaza

M O A B

Arnon Gorge

Ziklag

En Rimmon

Beth Pelet

Jeshua

Beersheba

Moladah

Jekabzeel

Negev

0   5 km.
0   5 miles

★ District capital

▲ *Athens: the Parthenon on the Acropolis, among other buildings, was under construction during the days of Ezra and Nehemiah (mid-fifth century BC).*

Jews to return to Judea (8:1 – 20). Those concessions included financial support, the right of the Judeans to govern their own affairs, the appointment of judges and civil magistrates, and the granting of tax exemptions for temple personnel. The major accomplishment of the second return was a spiritual rebuilding of the Judeans — including the dissolution of racial intermarriages (ch. 9).

In 446 BC the dilapidated state of Jerusalem's defenses came to the attention of a pious Jew named Nehemiah, who served in the Persian court as cupbearer to King Artaxerxes. The king endorsed Nehemiah's request to supervise the rebuilding of the walls of Jerusalem. Nehemiah left for Judea in 445 BC. He surveyed the condition of the defenses of the city (Neh 2:11 – 16) and then rallied the Judeans behind him.

Despite opposition, they rebuilt the walls in fifty-two days (2:17 – 7:3).

Ezra and Nehemiah then led the people in a time of spiritual renewal that ultimately resulted in the signing of a covenant document whereby the people committed themselves to govern their lives in accordance with the law of Moses. After serving for twelve years as governor (Neh 5:14; 13:6), Nehemiah returned to the Persian court in 433/432 BC. Later he served a second term as governor of Judah.

From the biblical and extrabiblical data we gain a fair understanding of Judah's place in the Persian Empire. It was a province (Yehud) in the satrapy "Beyond the River." To the north was the province of Samaria, whose governor, Sanballat, came into direct conflict with Nehemiah. To the east was the province of Gilead. The governor there was Tobiah the Ammonite, who had close ties with Eliashib, the priest in Jerusalem (Neh 13:4 – 7). On the west the "people of Ashdod" are mentioned among Nehemiah's opponents (4:7).

Besides Sanballat and Tobiah, "Geshem the Arab" (Neh 6:1) joined in the hostilities against Nehemiah. Geshem's kingdom was located in (northern) Sinai; he was likely involved in controlling the overland transport of luxury goods (e.g., gold, frankincense, myrrh, pearls, and spices) that passed through his territory on their way from Arabia to urban centers in the north and west.

As for Judah itself, Jewish settlements were established in the old Benjamite territory north, east, and west of Jerusalem (Ezra 2:21 – 35; Neh 7:26 – 38; 11:31 – 36). On the west Jews settled in the Shephelah (Neh 11:29 – 30), while in the east Judean territory stretched to the banks of the Jordan River. In the southern Hill Country of Judah Jews resided in Kiriath Arba (= Hebron) and further south in the Negev (11:25 – 28).

At the end of the fifth century BC the biblical record becomes silent. From papyrus documents found at Elephan-

Mark Connally

▲ *Tobiad family tomb at Tyre of Transjordan (Neh 2, 4, 6, 13) plus details of Tobiad inscription right of door*

tine, a site in southern Egypt (see *ANET*, 491 – 92; map p. 92), we know that a Jewish colony had been established there. The Jews had built a temple for worshiping Yahweh there, but it was destroyed around 410 BC. Also, we know the Jews of Elephantine on occasion looked to Jerusalem for guidance, for they corresponded with the governor of Yehud.

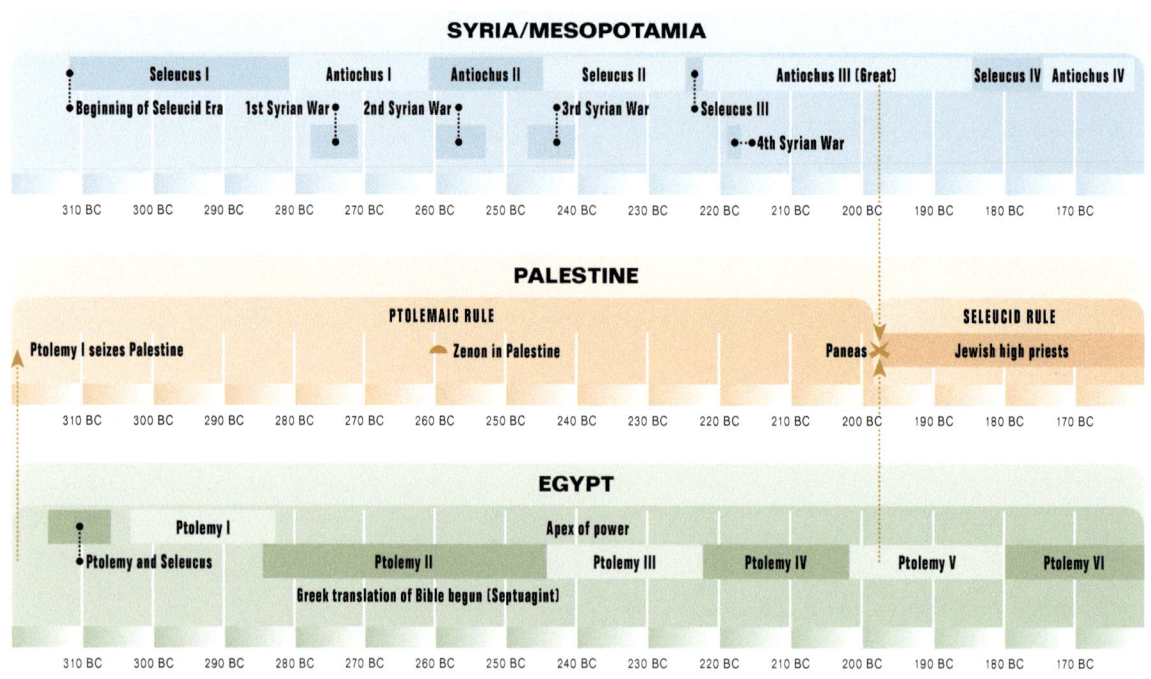

SYRIA/MESOPOTAMIA

Seleucus I · Antiochus I · Antiochus II · Seleucus II · Antiochus III (Great) · Seleucus IV · Antiochus IV

Beginning of Seleucid Era · 1st Syrian War · 2nd Syrian War · 3rd Syrian War · Seleucus III · 4th Syrian War

310 BC 300 BC 290 BC 280 BC 270 BC 260 BC 250 BC 240 BC 230 BC 220 BC 210 BC 200 BC 190 BC 180 BC 170 BC

PALESTINE

PTOLEMAIC RULE · SELEUCID RULE

Ptolemy I seizes Palestine · Zenon in Palestine · Paneas · Jewish high priests

310 BC 300 BC 290 BC 280 BC 270 BC 260 BC 250 BC 240 BC 230 BC 220 BC 210 BC 200 BC 190 BC 180 BC 170 BC

EGYPT

Ptolemy I · Apex of power

Ptolemy and Seleucus · Ptolemy II · Ptolemy III · Ptolemy IV · Ptolemy V · Ptolemy VI

Greek translation of Bible begun (Septuagint)

310 BC 300 BC 290 BC 280 BC 270 BC 260 BC 250 BC 240 BC 230 BC 220 BC 210 BC 200 BC 190 BC 180 BC 170 BC

THE ARRIVAL OF THE GREEKS

▼ *Cilician Gates, though which Darius III and Alexander the Great passed as they entered/exited Asia Minor.*

During the sixth through fourth centuries BC, the Greeks began to challenge Persian supremacy in Asia Minor. The Persians, who had invaded Greece, were kept at bay by the Greek forces, which defeated them, first at Marathon in 490 BC and then at Salamis, 480 BC. Through the rest of the fifth century and much of the fourth, the Greeks encouraged anti-Persian activities, but the extent of the territory under Persian rule remained virtually unchanged.

By the middle of the fourth century BC Philip II of Macedon solidified his position as ruler of Macedonia. During his life he worked toward two great goals: to unify the Greek city-states under his rule and to overthrow the Persians. In doing so he built his Macedonian army into a small but formidable fighting force. Unfortunately, Philip was assassinated in 336 BC, after completing plans to invade Asia Minor.

Philip's son Alexander ("the Great") was well positioned to carry out his father's dreams. Although only twenty years old, Alexander had been educated by Aristotle and had already led campaigns on behalf of his father. He began his invasion of Asia Minor by crossing the Dardanelles in 334 BC. He and his general Parmenio crushed all resistance throughout Asia Minor.

Alexander proceeded southeast to Tarsus via the Cilician Gates. Continuing around the

London, British Museum

▲ *Coin with a representation of Alexander the Great*

northeastern corner of the Mediterranean Sea, he passed through the Syrian Gates before realizing that Darius, the Persian monarch, had assembled a large army at Issus behind him. Retracing his steps he defeated Darius III at the battle of Issus in 333 BC.

Alexander then marched south along the coast of the Levant, securing or seizing Aradus, Byblos, Sidon, Tyre, Gaza, and other cities. After he crossed northern Sinai, Egypt submitted to his rule, and in the winter of 332–331 BC, he

Historical Section: *The Arrival of the Greeks* | 97

founded the city of Alexandria. This city became the capital of Egypt and a leading commercial and intellectual center of the world.

Alexander left Ptolemy in charge of Egypt and went north through the Levant. His treatment of the Judeans is not known, but most likely the Jews did not interfere with his advance along the coast. The Samaritans, however, murdered their governor Andromachus. In retaliation, Alexander destroyed Samaria and resettled it with Macedonian veterans.

Heading north and then east, Alexander passed through northern Mesopotamia. To the east of the Tigris, the Persians were soundly defeated. Darius fled further east but was murdered by Bessus, the satrap of Bactria. This ended nearly 200 years of Persian rule.

THE EMPIRE OF ALEXANDER THE GREAT

0   200 km.

0   200 miles

Alexander continued eastward to the area of modern Afghanistan and Kashmir, and then headed south into the Indus Valley (in modern Pakistan). From there he embarked on a difficult march west, through the deserts and mountains of southern Persia (modern Iran) toward Babylon. In 323 BC, at age thirty-two, Alexander suddenly died. But the Near East was radically changed with the arrival of Greek language and culture.

After Alexander's death a number of his officers quarreled over the conquered lands. Eventually, Antipater and Cassander were established in Macedonia and Greece; Lysimachus in Thrace and Asia Minor; Seleucus I in Syria, Mesopotamia, and Persia (all the way to the Indus River); and Ptolemy I in Egypt and Palestine.

During the reign of Ptolemy I (304–282 BC), the famous library and "museum" of Alexandria were established. His

political and military fortunes were varied. At one time he was able to extend his rule into southern Turkey and even toward the Greek mainland. These territories would pass in and out of Ptolemaic hands during the third century.

To the north of Palestine, Seleucus I (312 – 280 BC) established his capital at Antioch on the Orontes River. The influence of the Seleucid state was so pervasive that the calendrical system used in the Near East for hundreds of years was reckoned from the beginning of his reign (now known as 312 BC).

The Ptolemies and Seleucids fought wars during the third century BC, with Palestine caught in the middle. In the main the Ptolemies were successful in defending and controlling their territory. Palestine was divided into a number of administrative units called hyparchies. It also served to provide Egypt with quality olive oil, wines, wood products, and, at times, slaves.

Not much is known about the hyparchy of Judea during the third century BC. It would appear that few changes occurred in its size or internal administration, whose chief Jewish official was the high priest.

The hyparchy of Samaria, north of Judea, was populated with Macedonian veterans. The Samaritans maintained

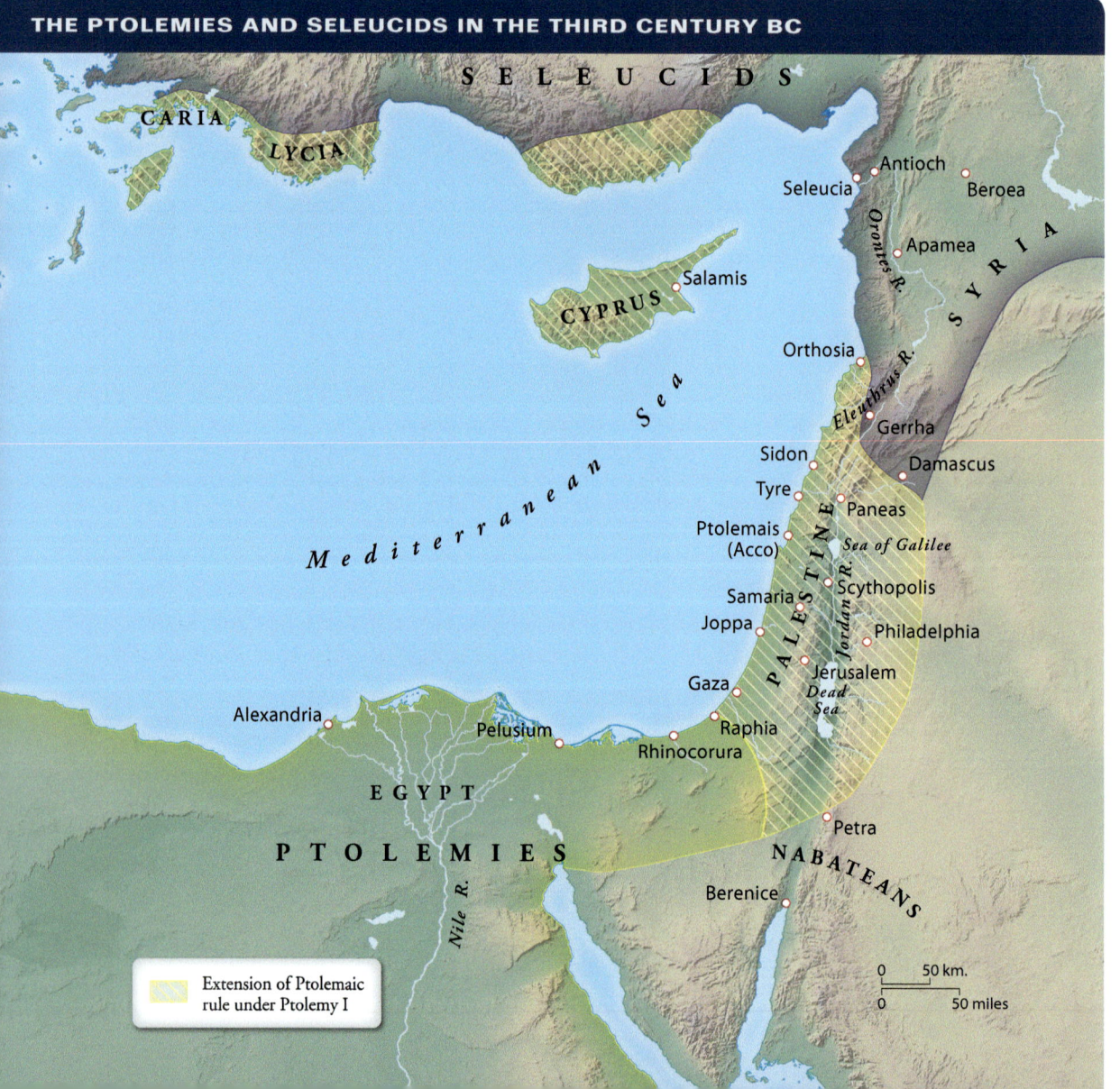

THE PTOLEMIES AND SELEUCIDS IN THE THIRD CENTURY BC

SELEUCIDS

CARIA

LYCIA

Antioch
Seleucia
Beroea
Apamea
Orontes R.
SYRIA

Salamis
CYPRUS

Orthosia
Eleuthrus R.
Gerrha

Sidon
Damascus
Tyre
Paneas
Ptolemais (Acco)
Sea of Galilee
Samaria
Scythopolis
Joppa
Jordan R.
PALESTINE
Philadelphia
Gaza
Jerusalem
Dead Sea

Mediterranean Sea

Alexandria
Pelusium
Raphia
Rhinocorura

EGYPT

PTOLEMIES

Petra

NABATEANS

Berenice

Nile R.

0   50 km.
0   50 miles

Extension of Ptolemaic rule under Ptolemy I

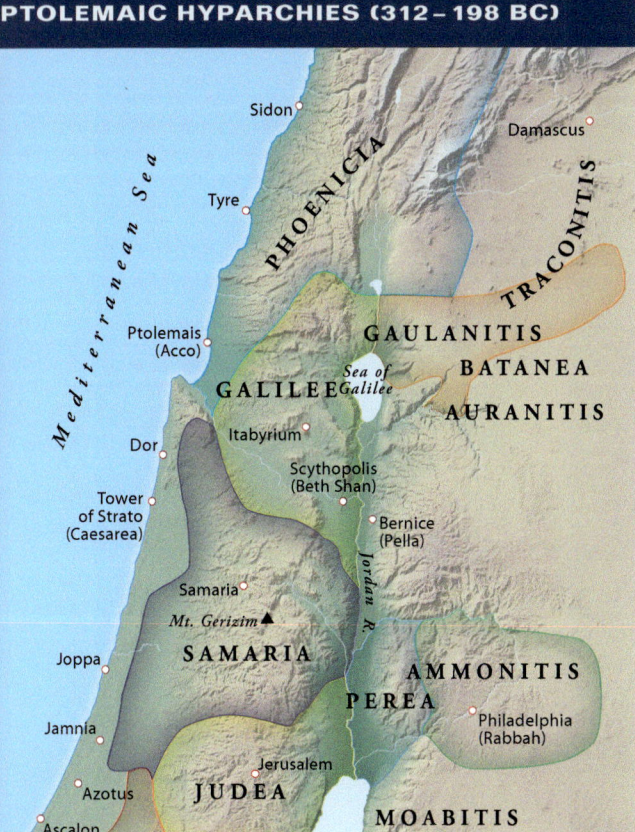

their religious and political institutions on and near Mount Gerizim. To the north of Samaria were royal estates in the Jezreel Valley, and to the north of these was the hyparchy of Galilee.

Coastal cities such as Tyre and Sidon had a high degree of independence. They supplied ships and sailors for the Ptolemaic navy and merchant fleet. Farther south, Acco, the port of Galilee, was one of the few cities to receive a dynastic name — Ptolemais. To the south was the Philistine Plain, and to the east of that was the hyparchy of Idumea.

When Ptolemy IV ascended the Egyptian throne and Antiochus III (223–187 BC) the Seleucid throne, the balance of power in the Levant began to shift in favor of the Seleu-

cids. In the Fourth Syrian War (221–217 BC) Antiochus pushed south into Galilee, the Jezreel Valley, and Transjordan, and advanced south to Gaza and Raphia, where he was defeated. But in 198 BC Antiochus III defeated the Ptolemaic general Scopas in the decisive battle at Paneas. Thus began a half century of Seleucid control of Judea (198–142 BC).

Initially, the Judeans prospered under the leadership of the Jewish high priest Onias III (198–174 BC). Because the Jewish population of Jerusalem had so readily received Antiochus III, they were granted special privileges, including the restoration of Jerusalem, limited tax exemptions, subsidies for the temple, and permission to live according to their ancestral laws (Josephus, *Ant.* 12.3.3, 4 [138–46]).

Mark Connally

▲ *Pleasure palace at Iraq el-Amir in Jordan from the Hellenistic Period*

In general the Seleucids encouraged the adoption of Greek language, culture, and customs, and they established new cities on the model of the Greek polis — where the adult male citizens met together to govern the affairs of the city. The Seleucids changed the Semitic names of cities to Greek ones. For example, Jerusalem became Antiochia. But these names lasted for only a brief period of time.

The Seleucids combined several Ptolemaic hyparchies to form a larger unit called an "eparchy." One of the largest of these was the eparchy of Samaria, whose governor resided in Samaria. During the latter part of his rule, Antiochus III experienced a series of defeats at the hands of the Romans, after which he relinquished control of much of Asia Minor and pledged to pay a heavy tribute to Rome.

Antiochus III (the Great) was succeeded by his son Seleucus IV Philopater, who maintained good relations with the Judeans — even presenting gifts to the temple in Jerusalem (2 Macc 3:3). At his death in 175 BC, Antiochus IV seized the throne. With the rise of Antiochus IV, Judea entered a critical phase in its history.

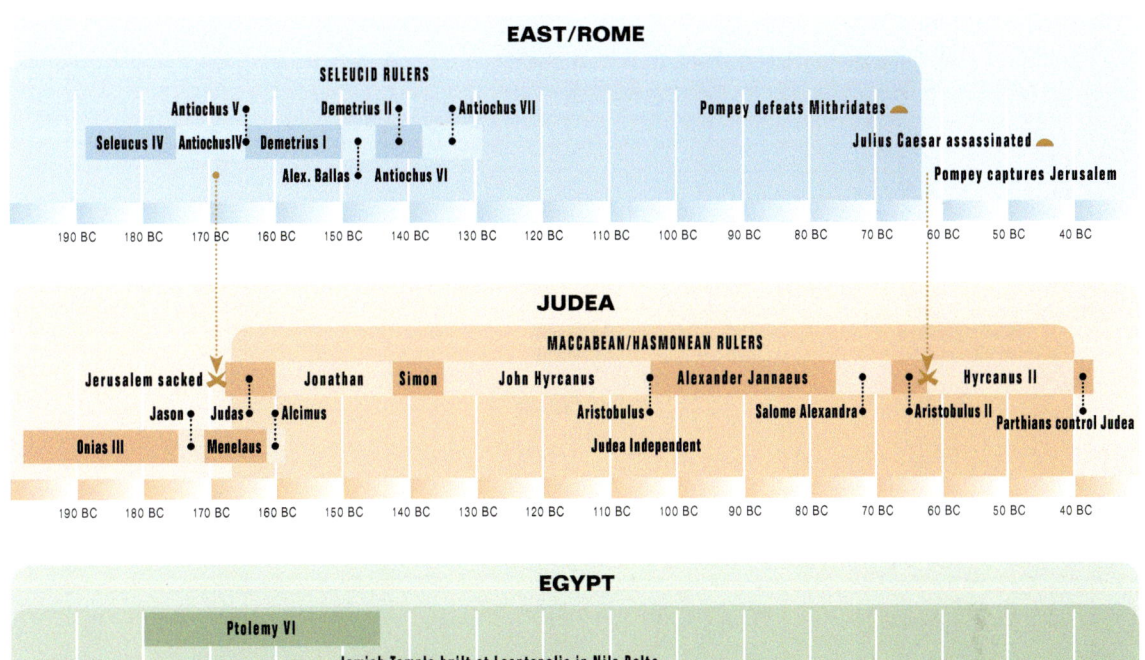

EAST/ROME

SELEUCID RULERS

			Antiochus V •		Demetrius II •		• Antiochus VII		Pompey defeats Mithridates		
Seleucus IV	• AntiochusIV•	Demetrius I							Julius Caesar assassinated	
				Alex. Ballas •	• Antiochus VI				Pompey captures Jerusalem	

190 BC 180 BC 170 BC 160 BC 150 BC 140 BC 130 BC 120 BC 110 BC 100 BC 90 BC 80 BC 70 BC 60 BC 50 BC 40 BC

JUDEA

MACCABEAN/HASMONEAN RULERS

Jerusalem sacked		Jonathan	Simon	John Hyrcanus		Alexander Jannaeus		Hyrcanus II	
Jason •	Judas•	• Alcimus			Aristobulus•		Salome Alexandra	• Aristobulus II	
Onias III	• Menelaus				Judea Independent				Parthians control Judea

190 BC 180 BC 170 BC 160 BC 150 BC 140 BC 130 BC 120 BC 110 BC 100 BC 90 BC 80 BC 70 BC 60 BC 50 BC 40 BC

EGYPT

Ptolemy VI

Jewish Temple built at Leontopolis in Nile Delta

190 BC 180 BC 170 BC 160 BC 150 BC 140 BC 130 BC 120 BC 110 BC 100 BC 90 BC 80 BC 70 BC 60 BC 50 BC 40 BC

THE MACCABEAN REVOLT AND THE HASMONEAN DYNASTY

▼ *Temple of Zeus at Gerasa (Jordan) — one of the Greco-Roman cities of the Decapolis. Originally built in the Hellenistic/Early Roman period.*

With the rise of Antiochus IV Epiphanes (175 – 163 BC), a chain of events began that culminated in the establishment of an independent Jewish state in 142 BC, which lasted until the Romans captured Jerusalem in 63 BC. These events had a direct influence on Jewish life and practice for the next two centuries.

Antiochus IV attempted to solidify his kingdom under the banner of Hellenism. Many Jews quickly adopted the new Hellenistic lifestyle (2 Macc 4), which meant breaking with their religious, cultural, and linguistic heritage. In Judea they gained more influence after the pious high priest Onias III (2 Macc 4:7) was deposed — the office going to the "highest bidder."

In 169 BC Jason, a former high priest, attempted to revolt against Antiochus IV. Antiochus responded by recapturing Jerusalem; thousands of Jews were killed or sold into slavery, the temple treasury was plundered, and Antiochus, a Gentile, entered the most sacred room of the temple, the Most Holy Place.

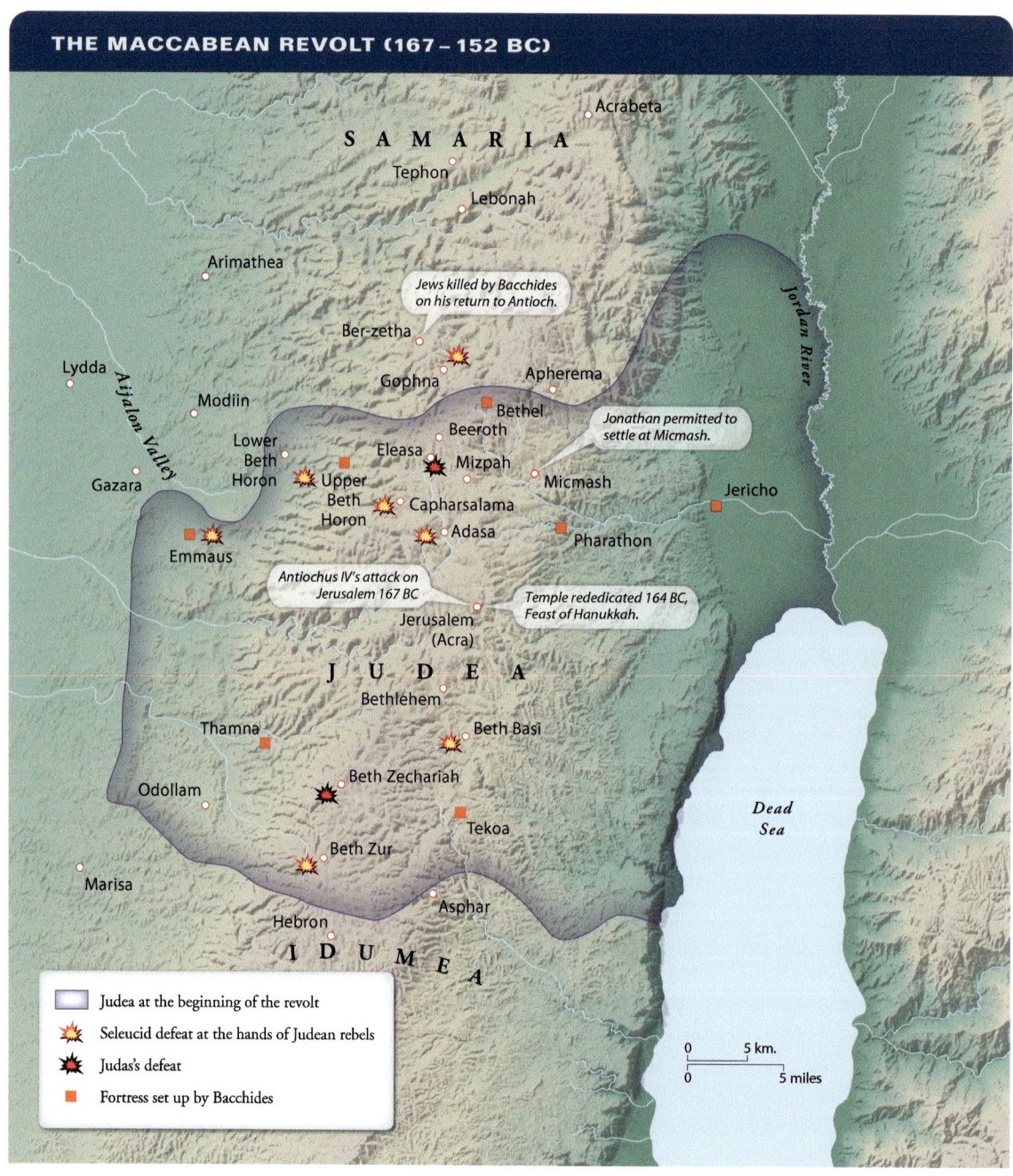

THE MACCABEAN REVOLT (167 – 152 BC)

SAMARIA

Acrabeta

Tephon

Lebonah

Arimathea

Jews killed by Bacchides on his return to Antioch.

Ber-zetha

Gophna

Lydda

Ajalon Valley

Modiin

Lower Beth Horon

Eleasa

Beeroth

Bethel

Apherema

Jonathan permitted to settle at Micmash.

Gazara

Upper Beth Horon

Mizpah

Capharsalama

Micmash

Jericho

Adasa

Emmaus

Pharathon

Antiochus IV's attack on Jerusalem 167 BC

Temple rededicated 164 BC, Feast of Hanukkah.

Jerusalem (Acra)

Jordan River

JUDEA

Bethlehem

Thamna

Beth Basi

Odollam

Beth Zechariah

Dead Sea

Tekoa

Beth Zur

Marisa

Asphar

Hebron

IDUMEA

Judea at the beginning of the revolt

Seleucid defeat at the hands of Judean rebels

Judas's defeat

Fortress set up by Bacchides

0   5 km.

0   5 miles

Antiochus IV tried to gain control of Egypt, but he was rebuffed by the Romans. He then decided to strengthen his kingdom by solidifying his position in Palestine. In 167 BC he dispatched troops to Jerusalem. In an attempt to Hellenize the population further, Jews were commanded to worship Zeus and other pagan deities, to burn their copies of the Torah, and to forsake the laws of their God (1 Macc 1:41 – 64). They were forbidden to observe the Sabbath, to celebrate their feasts, to sacrifice to God, and to circumcise their children. Portions of the walls of Jerusalem were torn down and a pagan citadel, called the Acra was constructed. The temple in Jerusalem was turned into a temple of Olympian Zeus, and on December 16, 167 BC, an unclean sacrifice was offered to Zeus (2 Macc 6:1 – 11).

Also in 167 BC, a delegate of Antiochus IV attempted to force Mattathias, a priest who lived in Modiin, to sacrifice to a pagan deity. Mattathias refused, but another Jew volunteered to perform the rite. Outraged, Mattathias killed both the Seleucid delegate and the errant Jew, and thus the Maccabean revolution began (1 Macc 2:1 – 48). The aged Mattathias soon died a natural death, leaving his five sons to carry the revolutionary torch (2:49 – 70).

The leader of the revolution was Judas, Mattathias's middle son, who was also called Maccabeus ("the hammerer"). Judas gained the support of the Hasidim, the "pious ones," who were true to ancient Jewish beliefs and practices. Judas and his followers went throughout the countryside, tearing down pagan altars and circumcising Jewish children.

In 165 BC the Seleucids, led by Lysias, assembled a large army at Emmaus. Judas mustered his warriors at the old tribal center of Mizpah in the Hill Country of Benjamin. Through a surprise attack, the Seleucids were defeated and retreated to the coast (1 Macc 3:27 – 4:25).

Lysias and Judas next clashed at Beth Zur, where Lysias was routed (1 Macc 4:26 – 35). Flushed with victory Judas marched to Jerusalem and recaptured the city, except for the Acra — which remained in the hands of the Hellenizers. The temple was cleansed, and on Chislev 25 (December 14), 164 BC, the temple was rededicated and proper Jewish sacrifices were resumed. Jews have commemorated this event as the Feast of Hanukkah — the Feast of Dedication (vv. 36 – 61).

After the death of Antiochus IV in 164 BC Judas and his brothers expanded their influence north into Galilee and south and southwest of Judah. But a large portion of the popu-

▲ Tombs of the "sons of Hezir" (left) and of "Zechariah" (center) in the Kidron Valley, Jerusalem. Note the Doric and Ionic columns and the pyramid-shaped roof, indicating Greek and Egyptian influence on the country.

▼ Mt. Gerizim: monumental staircase from the second century BC that led up to the Samaritan temple on the top of Mt. Gerizim. The temple was destroyed by John Hyrcanus in 110 BC (compare John 4:20).

lace desired closer ties with the Seleucids, and they appealed to Antiochus V for help. Lysias again marched to Beth Zur. He defeated Judas at Beth Zechariah and lay siege to Jerusalem.

PALESTINE OF THE MACCABEES AND THE HASMONEAN DYNASTY

Legend:
- Judea at the beginning of the revolt
- Additions of Jonathan, 160–142 BC
- Additions of Simon, 142–134 BC
- Additions of Hyrcanus I, 134–104 BC
- Additions of Aristobulus I, 104–103 BC
- Additions of Alexander Jannaeus, 103–76 BC
- Kingdom of Alexander Jannaeus

Sidon
Damascus
COELE-SYRIA
PHOENICIA
Tyre
Dan (Antiochia)
Paneas
Cadasa
Seleucia
Hazor
Bascama
Ptolemais
Bethsaida
Gamala
Gennesaret
Dathema
Taricheae
Sea of Galilee
Arbela
Hippus
GALILEE
Philoteria
Sepphoris
Mt. Carmel
Dora
Jezreel Valley
GALAADITIS
Strato's Tower
Scythopolis
Pella
SAMARIA
Gerasa
Samaria
Ammathus
Mt. Gerizim
Shechem
Apollonia
Acrabeta
Alexandrium
Joppa
Arimathea
Apherema
Jordan R.
Lydda
PEREA
Gadora
Docus
Philadelphia
Jamnia
Gazara
JUDEA
Jericho
Esbus
Azotus
Accaron
Samaga
Jerusalem
Hyrcania
Medeba
Ascalon
Herodium
Anthedon
Beth Zur
Machaerus
Gaza
Marisa
Adora
Hebron
Orda
En Gedi
Dead Sea
Gerar
IDUMEA
Masada
MOABITIS
Raphia
Beersheba
NABATEANS
Rhinocorura
Malatha
Mediterranean Sea

0   10 km.
0   10 miles

Wadi el-Arish

Petra

But because of internal problems he had to return to Antioch and so made peace with Judas, guaranteeing the Jews religious freedom (1 Macc 6:55 – 63), though he demanded that the walls of Jerusalem be torn down. Thus at least the religious gains of the revolt were preserved.

The next eighteen years (160 – 142 BC) were unstable, with battles, intrigue, and changing loyalties among the Hellenists, the Hasidim, the Maccabees (led now by Jonathan), and the Seleucid rulers. Eventually, after Jonathan was killed, his brother Simon allied himself the Seleucid Demetrius II, who sent him a letter confirming Judea's complete independence. Thus 142 BC marked the official independence of the Judean state — the first time that it had been officially free from foreign domination since 586 BC, when Jerusalem had fallen to the Babylonians. The Jews conferred on Simon the position of governor and high priest "for ever, until a trustworthy prophet should arise" (1 Macc 14:25 – 43). Thus with Simon, the Hasmonean dynasty (142 – 63 BC) began.

When Antiochus VII was killed in battle, strong Seleucid rule effectively ceased, and the Judeans engaged in expansionist activities. In 128 BC John Hyrcanus, a son of Simon, seized parts of Transjordan. In the same year he attacked the Samaritans, who had been harassing the Jews, and in 110 BC he destroyed their temple on Mount Gerizim. John Hyrcanus also established an alliance with Rome, and Rome confirmed his independence. In 125 BC he was able to move against Idumea, forcing them to convert to Judaism.

After his long and successful reign (135 – 104 BC), John Hyrcanus was replaced by his son Aristobulus I, who ruled for only one year (104 – 103 BC). Upon his death, Aristobulus's wife, Salome Alexandra, released his three brothers from prison and appointed one of them, Alexander Jannaeus, as king and high priest. She in turn married him, in spite of the fact that the high priest was supposed to marry only a virgin. Alexander Jannaeus's long reign (103 – 76 BC) was the high point of Hasmonean power, though it was marred by internal discord.

During this time conflict between the Sadducees and Pharisees came to a head. Jannaeus sided with the Sadducees and on occasion went out of his way to offend the Pharisees. For example, during a celebration of the Feast of Tabernacles, instead of pouring the sacred water on the altar, he poured it on his feet. The worshipers at the temple responded by pelting him with lemons; Alexander, in turn, responded by massa-

▲ *Qumran Cave IV: numerous Dead Sea Scroll fragments were found in this cave.*

cring some 6,000 Jews. Using foreign mercenaries, Alexander Jannaeus fought with his own countrymen over a six-year period, with the result that almost 50,000 Jews were eventually killed in the conflict. At one point he crucified 800 Pharisees. This tragic incident illustrates the fact that the conflict between the Pharisees and Sadducees, evident on the pages of the New Testament, was more than a theological dispute; its history was punctuated with literal life-and-death matters.

Alexander Jannaeus realized that the Pharisees enjoyed popular support, and on his deathbed he instructed his wife, Salome Alexandra, to make peace with them. After his death, Salome Alexandra assumed civil rule (76 – 67 BC). She appointed Alexander Jannaeus's son Hyrcanus II to be high priest. She made peace with the Pharisees, and her rule was characterized by peace. However, friction developed when Aristobulus II, the youngest son of Alexander Jannaeus, wanted to be appointed high priest.

After Salome Alexandra's death, the two Hasmonean brothers continued their conflict. It was only resolved by Pompey, who decided to put an end to the intrigue. In 63 BC, after a three-month siege of the temple area, Roman soldiers entered the temple area and killed 12,000 Jews. Pompey reinstalled Hyrcanus II as high priest (63 – 40 BC), though with much more limited powers; Judea and Jerusalem were now firmly under Roman control. Thus in 63 BC the Hasmonean state, independent since 142 BC, officially ceased to exist.

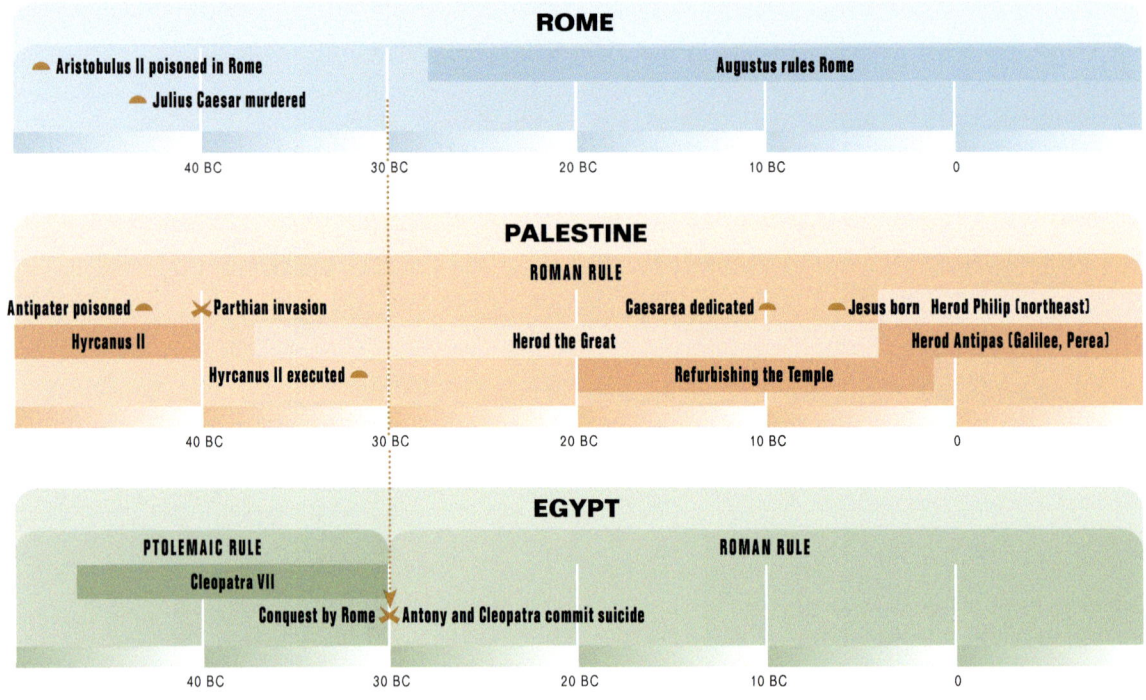

ROME

- Aristobulus II poisoned in Rome
- Julius Caesar murdered

Augustus rules Rome

| 40 BC | 30 BC | 20 BC | 10 BC | 0 |

PALESTINE

ROMAN RULE

Antipater poisoned — ✕ Parthian invasion

Caesarea dedicated —

— Jesus born Herod Philip (northeast)

Hyrcanus II

Herod the Great

Herod Antipas (Galilee, Perea)

Hyrcanus II executed —

Refurbishing the Temple

| 40 BC | 30 BC | 20 BC | 10 BC | 0 |

EGYPT

PTOLEMAIC RULE

ROMAN RULE

Cleopatra VII

Conquest by Rome ✕ Antony and Cleopatra commit suicide

| 40 BC | 30 BC | 20 BC | 10 BC | 0 |

EARLY ROMAN RULE IN PALESTINE

W hen Pompey withdrew from the Near East, he left behind a proconsul to govern the province of Syria, of which Judea was a part. Thus, the sphere of Jewish control was greatly reduced. Along the Mediterranean coast, cities were granted autonomous status directly under the proconsul. Even the Jewish port of Joppa was detached from Judea.

Greco-Roman cities to the east of the Jordan River, along with Scythopolis to the west, were also freed from Jewish control, and their Gentile populations (people exiled by the Maccabees and Hasmoneans) were encouraged to return. Some of these cities banded together into a league called the Decapolis

("ten cities"). Jewish territory was limited to Judea proper, eastern Idumea, Perea, and a portion of Galilee.

The Romans appointed Hyrcanus II (63 – 40 BC) as high priest and left him in charge of Jewish affairs. During this time the Roman Empire was racked by civil strife, which began when Julius Caesar crossed the Rubicon in 49 BC. By 48 BC Caesar was gaining the upper hand against Pompey, pursuing him to Egypt. Hyrcanus II supported Caesar and instructed the Jews of Egypt to do the same. In addition, Antipater the Idumean, the power behind Hyrcanus, also supported Caesar. As a result, Caesar confirmed Hyrcanus II as high priest and ethnarch and appointed Antipater as procurator. Antipater appointed his sons Phasael and Herod as governors in Jerusalem and Galilee.

In 44 BC Julius Caesar was murdered, and civil war resumed in Rome. In 42 BC, Antony became master of Roman holdings in Asia. But in 40 BC the Parthians invaded Palestine and installed Antigonus II, a Hasmonean, as king and high priest in Jerusalem; Hyrcanus II was taken to Parthia as a prisoner. Herod fled to the fortress of Masada but eventually made

Tyre

Paneas

ITUREANS

L. Semechonitis

PHOENICIA

GAULANITIS

Ptolemais

GALILEE

Sea of Galilee

Sepphoris

Hippus

Dium

Philoteria

Gadara

Abila

Dora

Valley of Esdraelon

D E C A P O L I S

Strato's Tower

Scythopolis

Pella

Gerasa

Mediterranean Sea

Samaria

Shechem

Mt. Gerizim ▲

SAMARITANS

Ammathus

Jordan R.

Apollonia

Alexandrium

Thamna

Philadelphia

Joppa

Lydda

Gophna

J U D E A

Jamnia

Emmaus

Jericho

P E R E A

Azotus

Jerusalem

Esbus

Ascalon

Medeba

Hebron

Machaerus

Gaza

Adora

Dead Sea

I D U M E A

Masada

Raphia

N A B A T E A N S

◼ Autonomous cities directly
under the proconsul

☐ Sphere of Jewish control

0   10 km.

0   10 miles

HEROD THE GREAT'S KINGDOM

Sidon

Damascus

Mediterranean Sea

Tyre

S Y R I A

Paneas

P H O E N I C I A

U L A T H A

Meroth

I T U R E A N S

Ptolemais

GALILEE

GAULANITIS

B A T A N E A

TRACONITIS

Gaba

Mt. Carmel

Tiberias

Sea of Galilee

Hippus

Dion

A U R A N I T I S

Sepphoris

Jezreel Valley

Gadara

Abila

Caesarea Maritima
(Strato's Tower)

Scythopolis

D E C A P O L I S

SAMARIA

Pella

Sebaste
(Samaria)

Gerasa

Mt. Gerizim

Ammathus

Alexandrium

Joppa

Antipatris

Phasaelis

P E R E A

Philadelphia

Jamnia

J U D E A

Jericho

Esbus

Emmaus

Jerusalem

Cypros

Azotus

Bethlehem

Hyrcania

Ascalon

Herodium

Callirrhoe

Betogabris

Tekoa

Anthedon

Dead Sea

Machaerus

Gaza

Hebron

Orhesa

I D U M E A

N A B A T E A N S

Beersheba

Masada

Malatha

Jordan R.

0   10 km.

0   10 miles

◆ Military colony founded by Herod

■ Herodian fortress

▢ Herod's kingdom at the start of his reign

▢ Additions to Herod's kingdom

his way to Rome, where he was warmly received by Octavian and Antony, who persuaded the senate to appoint him king of Judea and to add Samaria and western Idumea to his realm.

From 40 BC until 37 BC Herod fought to gain control of the territory that the Romans had granted him. He began by capturing the port city of Joppa. Then he marched against Antigonus II in Jerusalem, though his first attempt to capture the city failed. During the winter of 39/38 BC he subdued Galilee. After strengthening his forces, he laid siege to Jerusalem in the winter of 38/37 BC and took the city the following summer.

From 37 BC to 25 BC, Herod consolidated his kingdom. Internally, he faced opposition from the Pharisees, from remnants of the Hasmonean family, and from portions of the populace and aristocracy. Herod, of Idumean descent, was never accepted by the Jewish population at large as a true Jew. In an attempt to legitimize his claim to the kingship, Herod married the Hasmonean Mariamne. Her mother, Alexandra, was able to secure a Hasmonean foothold in the religious-political structure of the government by having her seventeen-year-old son Aristobulus appointed as high priest.

The people saw him as a legitimate Jewish replacement for Herod. Enraged by this, Herod arranged for some of his friends to hold the young Aristobulus under water too long while they were swimming in one of the pools in Jericho. In spite of Herod's feigned grief over the "accidental" death of Aristobulus, it was well known that he, in fact, was the instigator of the deed. Herod also eventually eliminated Mariamne, her mother, and the aged Hyrcanus. By 25 BC, most of the internal threats to his kingship had been removed.

Externally, Herod faced a formidable threat from Cleopatra in Egypt, who wanted to revive the Ptolemaic Empire into Palestine and Arabia. Antony, her lover and the master of the east, agreed, and in 35 BC granted her large portions of Herod's territory and of Arabia. But Antony was defeated by Octavian in 31 BC at Actium, and he and Cleopatra committed suicide rather than face the wrath of Rome.

While Antony and Cleopatra were losing power, Herod skillfully changed his allegiance from Antony to Octavian so that when the latter emerged victorious, Herod received back the territories and cities that he had lost to Cleopatra. Thus he emerged from the crisis stronger than ever. From 25 to 14 BC, Herod added further territory to his kingdom.

▲ *Model of the temple in Jerusalem that Herod the Great refurbished and which was standing in the days of Jesus — up until its destruction by the Romans in AD 70.*

Herod then took a number of measures to secure his kingdom. He established at least two military colonies: one at Gaba and one at Esbus (= OT Heshbon). He also built or rebuilt a string of fortresses throughout his kingdom, which he used to control nearby territory and which served as places of security if Herod had to flee (e.g., Masada) or as prisons.

Herod also sought to neutralize the potential threat from the Jewish population by building and rebuilding cities along Greco-Roman lines and settling Gentiles in them (e.g., Samaria, which he renamed Sebaste, the Greek name for Augustus, the emperor). Since Herod now ruled most of the Mediterranean coastline, he moved to create a secure port for himself from which he could maintain constant contact with Rome and export grain crops to Rome. He chose a small landing called Strato's Tower, just south of Mount Carmel.

Strato's Tower was well situated, for an easy pass through the Mount Carmel range connected it with the Jezreel Valley and the rich agricultural areas northeast of the Sea of Galilee. There Herod built Caesarea Maritima, naming it after the emperor. He brought in fresh water, via tunnels and aqueducts, from springs located at the foot of Mount Carmel, and he built a huge port and other magnificent public buildings. Caesarea

BRITANNIA

BELGICA

LUGDUNENSIS

Oceanus
Atlanticus

NORICUM

RAETIA

AQUITANIA

ALPES
GALLIA
NARBONENSIS
VENETIA

PANNONIA

DACIA

TARRACONENSIS

GALLAECIA

LUSITANIA

Rubicon R.

ITALIA

ILLYRICUM
(DALMATIA)

MOESIA

Pontus Euxinus

Mare Caspium

CORSICA

Rome

BAETICA

SARDINIA

MACEDONIA

THRACIA

BITHYNIA

ARMENIA

Pergamum

GALATIA

CAPPADOCIA

COMMAGENE

PARTHIA

Actium

ASIA

LYCAONIA

MAURETANIA

NUMIDIA

AFRICA PROCONSULARIS

Mare   Internum

ACHAIA

PAMPHYLIA

CILICIA

LYCIA

CRETA

SYRIA

PHOENICIA

CYRENAICA

Alexandria

Pelusium

JUDAEA

Red Sea

AEGYPTUS

- ▢ Under Roman control in 100 BC
- ▢ Under Roman control at the time of Julius Caesar's death, 44 BC
- ▢ Extent of direct Roman rule at the death of Augustus, AD 14
- ▢ Area acquired after Augustus till AD c. 150

0   300 km.

0   300 miles

▼ *Herodium: interior courtyard of Herod the Great's pleasure palace. Note large circular eastern tower. Apartments once ringed the area.*

continued to expand in importance and soon became the capital of the country, a position it held for almost 600 years.

Herod lavished much attention and energy on Jerusalem. In 20 BC, he began refurbishing the temple area (see p. 146). Always thinking security, Herod strengthened the Antonia fortress, which overlooked all of the temple precincts. For himself he built a magnificent palace on the western hill and fortified the approach to it from the north by constructing three huge towers, named Hippicus (after a friend), Phasael (after his brother), and Mariamne (after his beloved wife, whom he had executed). The massive base of one of these towers still remains in the present-day citadel complex just southeast of the Jaffa Gate.

During the final period of Herod's rule (15 BC – 4 BC), the major concern was that of succession. During this time of intrigue, plots, slander, and duplicity, Herod drew up at least six wills, naming first one and then another of his sons as his successor. When Magi from the east appeared in Jerusalem asking, "Where is the one who has been born king of the

▲ *Herodium: seven miles south of Jerusalem. A combination pleasure palace, fort, and mausoleum built by Herod the Great.*

Jews?"(Matt 2:2) it is no wonder that "when King Herod heard this he was disturbed, and all Jerusalem with him" (2:3). His slaughter of the baby boys of Bethlehem to remove a possible threat to his throne (vv. 16 – 18) was certainly in keeping with his character.

Herod's health began to deteriorate rapidly, and he died in Jericho in the spring of 4 BC. Even though the Jewish popu-lation rejoiced at his death, his family and soldiers gave him a lavish funeral, carrying his body with great pomp in a jewel-studded gold coffin from Jericho to his mausoleum at the Herodium. He left behind a kingdom that was economically and materially prosperous. But he had also ruled using fear and terror and thus left behind much dissatisfaction. To some he was Herod the Great, to others, Herod the Despicable.

▼ *Herodium: Recently discovered foundation monument of the tomb of Herod the Great*

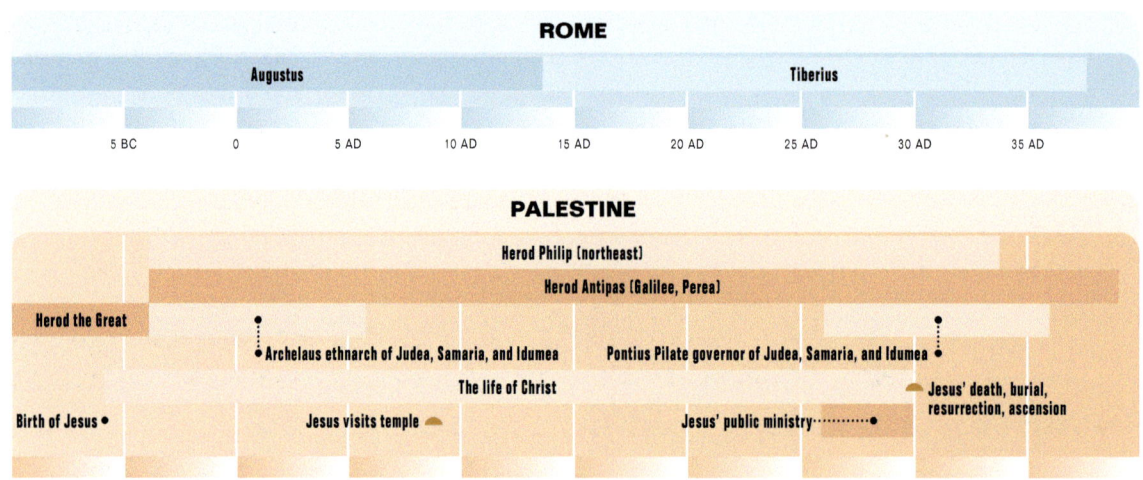

ROME

| Augustus | | | | Tiberius | | | | |

| 5 BC | 0 | 5 AD | 10 AD | 15 AD | 20 AD | 25 AD | 30 AD | 35 AD |

PALESTINE

Herod Philip (northeast)

Herod Antipas (Galilee, Perea)

Herod the Great

Archelaus ethnarch of Judea, Samaria, and Idumea   Pontius Pilate governor of Judea, Samaria, and Idumea

The life of Christ

Jesus' death, burial, resurrection, ascension

Birth of Jesus •   Jesus visits temple   Jesus' public ministry

| 5 BC | 0 | 5 AD | 10 AD | 15 AD | 20 AD | 25 AD | 30 AD | 35 AD |

THE LIFE OF CHRIST

In his sixth and final will Herod designated Archelaus as king of Idumea, Judea, and Samaria; Antipas as ruler in Galilee and Perea; and Philip as governor of the lands northeast of the Sea of Galilee. However, the Romans did not give Archelaus the title of king but rather "ethnarch" (meaning "ruler of the nation"). Archelaus's ten-year rule (4 BC – AD 6) was brutal. It is little wonder that when Mary, Joseph, and the baby Jesus returned from Egypt, they avoided returning to Judea, for they heard that Archelaus was ruling in place of his father (Matt 2:19 – 23). Instead, they proceeded to Galilee and settled in the village of Nazareth.

Herod Antipas (4 BC – AD 39) ruled over both Galilee and Perea. Each of these territories had a large number of Jews. The area northwest of the Sea of Galilee was higher in elevation and was called Upper Galilee. To the south, Lower Galilee was much more open to outside influence, and its broad, spacious valleys provided good land for growing grain crops.

As Jesus was growing up, Antipas was constructing his new capital at Sepphoris (3 BC – AD 10), which may have had a population of 5,000. This city overlooked valuable farmland

▲ *Nazareth: Church of the Annunciation surrounded by the hills rising above Nazareth*
▼ *A full-scale model of the excavated Galilee boat. This multi-purpose boat could carry about 15 persons total.*

and was close to an important east – west route that connected the cities of the area with the port of Ptolemais.

Jesus was raised in the small village of Nazareth, only 3.5 miles southeast of Sepphoris. Although Nazareth itself was small, its residents probably came into contact with caravans and Greek-speaking Gentile traders who passed through Sepphoris on the north or the Esdraelon Valley (= OT Jezreel Valley) on the south.

When Jesus began to minister at about age 30, he spent much more time in Lower than in Upper Galilee (map p. 116). He ministered at Cana of Galilee: turning water into wine and healing the son of a Roman official (John 2:1 – 11; 4:43 – 54). Two sites have been suggested for Cana: Khirbet Qana (8 miles north of Nazareth) and Kafr Kana (4 mi. northeast of Nazareth).

It is about 12 miles from Cana to the Sea of Galilee, about a six-hour walk. There, along the northern shore, Jesus spent much of his public ministry. The largest city on the lake was the newly built city of Tiberias, which Herod Antipas made his capital (AD 18 – 22; Josephus, *Ant.* 18.2.3 [36 – 38]). In John 6:1 and 21:1, the Sea of Galilee is called (lit.) "the Sea of Tiberias," and on one occasion, boats from Tiberias arrived with passengers wanting to see Jesus (6:23).

Four miles to the north of Tiberias, on the western shore of the sea, is the probable site of Magadan (Matt 15:39; Mark 8:10 calls it Dalmanutha). Jesus visited it after feeding the 4,000 on the other side of the lake. Proceeding from Magadan 6 miles

in a clockwise direction around the north shore of the Sea of Galilee is Capernaum. Apart from Jerusalem, this is the most important town mentioned in the Gospels, for here Jesus established his headquarters for the major portion of his public ministry. Several of his disciples were from Capernaum (Mark 1:21, 29). Fishing was probably its major occupation.

Capernaum sat astride a branch of the international route that ran from the Mediterranean Sea to Transjordan

THE DIVISION OF HEROD'S KINGDOM

Legend:
- ◆ Cities of the Decapolis (Pliny)
- ☐ Territory under Antipas
- ☐ Territory under Philip
- ☐ Territory under Procurator of Judea
- ☐ Territory under the Proconsul of Syria

and Damascus, and a custom station was located there, likely staffed by Matthew (Matt 9:9). The town was important enough for a Roman centurion and his troops to be stationed there (8:5 – 9). In Capernaum, Jesus healed many people, including the servant of the centurion (Matt 8:5 – 13), the paralytic who was let down through the roof of a dwelling (Mark 2:1 – 12), Peter's mother-in-law (1:29 – 31), and a royal official's son (John 4:46).

The Franciscans, who now own much of the site of Capernaum, have excavated a beautiful white limestone synagogue that dates from the sixth century AD; underneath it they have discovered the massive foundation walls of a black basalt synagogue that preceded it. This earlier synagogue probably dates back to the days of Jesus and was the one in which he preached while at Capernaum. Early Christian presence at the site is evidenced by the remains of several churches that were built over a house, thought to have been the house of Peter.

Although it is difficult to pinpoint the exact location of many of Jesus' activities in the neighboring countryside, by the fourth century Christian tradition had localized the site of the Sermon on the Mount (Matt 5 – 7), the feeding of the 5,000 (14:13 – 21), and the appearance of the resurrected Lord

JESUS IN GALILEE

▲ Black basalt foundation wall of an earlier synagogue upon which the rebuilt limestone walls of the sixth-century synagogue at Capernaum are visible.

▼ View of the Plain of Gennesaret and the Arbel Cliffs from the Mount of Beatitudes

▲ "Sowers Cove" (Matt 13:1–2) on the northwestern shore of the Sea of Galilee

to his disciples (John 21) near the place of seven springs — Heptapegon (Tabgha). This area may indeed have been the site of these events, although the feeding of the 5,000 probably occurred northeast of the Sea of Galilee. Between Capernaum and Tabgha is a small bay (called Sower's Cove) on the seashore in the shape of a natural theater that may have been the

spot where Jesus spoke "many things in parables" from a boat (Matt 13:2–3).

Another important village Jesus visited is Bethsaida. A possible location is the mound called et-Tell, located east of the Jordan River about 1.5 miles before it enters the Sea of Galilee. This city was built by Philip, the son of Herod the Great, who named it Julias, after Julia, the daughter of Augustus. However, there was another Bethsaida "in Galilee" (John 12:21). The latter has been tentatively identified with Araj, located close to the shore of the Sea of Galilee. Bethsaida was the early home of Peter, Andrew, and Philip (1:44; 12:21). There a blind man was healed (Mark 8:22–26), and in a nearby deserted place Jesus fed 4,000 people.

To the northeast of Bethsaida lay Philip's territory. In the first century, most of this territory was settled by Gentiles, and Jesus does not appear to have spent much time there. However, on at least one occasion he traveled with his disciples to the vicinity of Caesarea Philippi, about 25 miles north of Bethsaida. There, at the headwaters of the Jordan, Herod the Great had built a white marble temple in honor of the emperor; and there his successor, Philip, built a large city that he named after the emperor — adding his own name to the title.

Philip made Caesarea Philippi the capital of his territory, and it must have been a thriving city, for it was situated

▲ Rock-cut sanctuary of Pan at Caesarea Philippi — near where Peter affirmed that Jesus was the Messiah (Matt. 16)

▼ Wilderness of Judah east of Jerusalem. Jesus fasted for forty days in this area and passed through it on his way from Jericho to Jerusalem.

along the road that led from Damascus to Tyre and Sidon. In this vicinity Peter made his "great confession," stating that he believed that Jesus was "the Messiah, the Son of the living God" (Matt 16:13 – 20). Soon afterward Jesus was transfigured in the presence of Peter, James, and John (Matt 17:1 – 8; Mark 9:2 – 8; Luke 9:28 – 36). It is possible that the transfiguration also occurred in this region, perhaps on Mount Hermon.

To the south of Philip's territory was a region that came to be known as the Decapolis — a group of ten Greco-Roman cities (hence the name Decapolis, meaning "ten cities"), though in later years it often included more than ten cities. On one occasion Jesus healed two demon-possessed men (Matt 8:28), one of whom went into the Decapolis to tell of all that Jesus had done for him (Mark 5:20). The placement of this miracle near Gadara (modern Umm Qeis; see Matt 8:28) is the most plausible location (though also see Mark 5:1, which refers to Gerasa, much further south), since it is only 6 miles southeast of the sea.

To the south and west of the Decapolis was the region called Perea. This is a shortened form of a Greek phrase that can be translated as "other side of the Jordan" or "regions across the Jordan." Herod Antipas received this territory and controlled both it and Galilee. Perea, Galilee, and Judea are called "the three Jewish provinces" in the Mishnah (written around AD 200).

Jesus ministered in Perea, since Luke 9:51 – 18:34 places a number of events there. In addition, John was baptizing "at Bethany on the other side of the Jordan" (John 1:28). This Bethany is difficult to locate precisely, but it may have been in the vicinity of Bethennabris or a spot closer to the Jordan. Later, the gospel writer notes that John was "baptizing at Aenon near Salim, because there was plenty of water" (3:23). This Aenon ("springs") is also difficult to identify, but the best location is in/near the Jordan Valley near Salim. This places John's activities in the Decapolis, just outside the reach of Herod Antipas (who had been angered by his preaching) and of Pilate (who might have considered him a revolutionary).

Jews living in Perea probably had close contact with Jerusalem, for they could cross the fords of the Jordan opposite Jericho and climb their way up to the Holy City. In Jesus' day the Romans controlled Jericho, and its aqueducts, plantations, fortresses, palaces, and pools were spread out over a large area. Jesus mentioned Jericho in the parable of the good Samaritan (Luke 10:25 – 37), and he passed through it on his way to Bethany to raise Lazarus from the dead (John 10:40 – 11:54). At Jericho two blind men (Matt 20:29 – 34), including Bartimaeus (Mark 10:46), were healed, and Jesus also dined there with Zacchaeus, the tax collector (Luke 19:1 – 10).

From Jericho a well-traveled road ran up to Jerusalem through the dry, chalky wilderness. After an uphill walk of

Sidon

S Y R I A

Damascus

T Y R E

Tyre

Mt. Hermon

U L A T H A

Caesarea
Philippi

Mediterranean Sea

G A U L A N I T I S

TRACONITIS

Gischala

G A L I L E E

Bethsaida

B A T A N E A

Ptolemais

Raphana

Capernaum

Cana

Sea of Galilee

Gergesa (Kursi)

Canatha

Kafr
Kana

Sepphoris

Tiberias

Hippus

Dion

Nazareth

AURANITIS

Esdraelon Valley

Mt. Tabor

Gadara

Dora

Abila

Edrei

Caesarea

Scythopolis

D E C A P O L I S

Pella

Ginae

Salim

S A M A R I A

Aenon

Geba

Wadi Farah

Gerasa

Mt. Ebal

Jordan R.

Sebaste
(Samaria)

Sychar
Shechem

Ammathus

Apollonia

Mt. Gerizim

Acrabeta

Coreae

Antipatris

Anuathu Borcaeus

Gadora

Joppa

Lebonah

Phasaelis

P E R E A

Lydda

Ephraim
(Ophrah)

Philadelphia

Bethel

Archelais

Jamnia

J U D E A

Bethennabris

Emmaus?
Nicopolis
(Imwas)

Jericho

Abila

Esbus

Emmaus?
Qaloniya
(Mozah)

Jerusalem

Bethany

Azutus

Ascalon

Bethlehem

Mesad
Hasidim
(Qumran)

Bethany, on the
other side of
the Jordan

Medeba

I D U M E A

Gaza

Hebron

Machaerus

*Dead
Sea*

En Gedi

◆	Cities of the Decapolis (Pliny)
—	Major routes

0   10 km.

0   10 miles

▲ *Church and olive grove at the traditional site of the Garden of Gethsemane on the western foot of the Mount of Olives*

eight to ten hours one approached the eastern slopes of the Mount of Olives. Here was the village of Bethany, the home of Mary, Martha, and Lazarus. Jesus often stayed there, and events such as the teaching of Mary, the raising of Lazarus, and the anointing with precious oil took place there. From Bethany/Bethphage Jesus mounted a colt and rode it into Jerusalem on Palm Sunday. During the final week of his life, he spent several days teaching in Jerusalem, but he seems to have returned to Bethany every night.

The territory of Judea stretched thirty-five miles north of Jerusalem. Early in Jesus' ministry, this was probably the area where he and his disciples "went out into the Judean countryside" (John 3:22). Late in his ministry, after raising Lazarus and learning of a plot on his life, he withdrew with his disciples to this same area, to a "village called Ephraim" (11:54).

North of Judea was the district of Samaria (map p. 119), which stretched to the village of Ginae. This district too was governed by the Roman official, Pontius Pilate. The district was named after the Old Testament city of Samaria (then called Sebaste), and the Samaritans dominated large portions of the area. An important route that ran through Samaria was used by some Jewish inhabitants of Galilee on their pilgrimages to and from Jerusalem (Josephus, *Ant.* 20.6.1 [118]). This portion probably took three days. Galileans heading south crossed the Valley of Esdraelon and entered Samaria at Ginae. Here, "along the border between Samaria and Galilee" (Luke 17:11), Jesus probably met and healed ten lepers, one of whom was a Samaritan (vv. 12 – 19).

Jewish pilgrims would then continue south from Ginae toward Shechem, and they may have spent the night in the area of Geba. It is doubtful they stayed in Samaritan or Gentile homes, so they presumably camped out in the open. From Geba, the pilgrims continued south, passing Mount Ebal and Mount Gerizim. They likely entered Jewish Judea before settling in for the night, possibly in the el-Lubban (= OT Lebonah) region. The last day of their journey took them into Jerusalem.

On one occasion Jesus, heading north, stopped at "Jacob's well" near the town of Sychar (modern Askar) at midday (John 4:4–6); it is about a half-day's journey north from the el-Lubban overnight stop to Sychar. There, near the foot of the Samaritan holy mountain, Mount Gerizim, he pointed the Samaritan woman to the real source of living water so that she, and others like her, could worship God in spirit and truth (vv. 4–42).

Only one event in the Gospels is placed west of Jerusalem. This was his appearance to the two disciples on the road to Emmaus (Luke 24:13–35). According to the best Greek manuscripts, Emmaus was 60 stadia (ca. 7 mi.) from Jerusalem. One possible site is near modern Qaloniya/Motza — a site 3.5 miles west of Jerusalem on the Roman road leading to Joppa (map p. 119). If so, the distance in Luke 24:13 is the distance from Jerusalem to Emmaus and back, i.e., the distance of a round trip.

Another possible site for biblical Emmaus is the city of Emmaus/Nicopolis. The name of the ancient city was preserved in the now-destroyed Arab village of Imwas, which overlooked the Aijalon Valley. But this site is about 19 miles from Jerusalem (but note that one important Greek manuscript reads "160 stadia" [= ca. 20 mi.]).

It was back in the Jerusalem area, on the Mount of Olives, that Jesus ascended into heaven (for Jesus' final days in Jerusalem, see p. 148). It is amazing to reflect on the worldwide significance of the message and work of this first-century itinerant Jewish prophet, especially when one considers that he only ministered for three or four years, that he left behind only a small band of loyal followers, and that his ministry was primarily confined to a rather small province of the Roman Empire. But the New Testament writers were anxious to establish that it was not through the might of Herod the Great or through the power of the Roman emperors, but through Jesus, that all of the nations of the earth will be blessed (Gen 12:3; Gal 3:6–15).

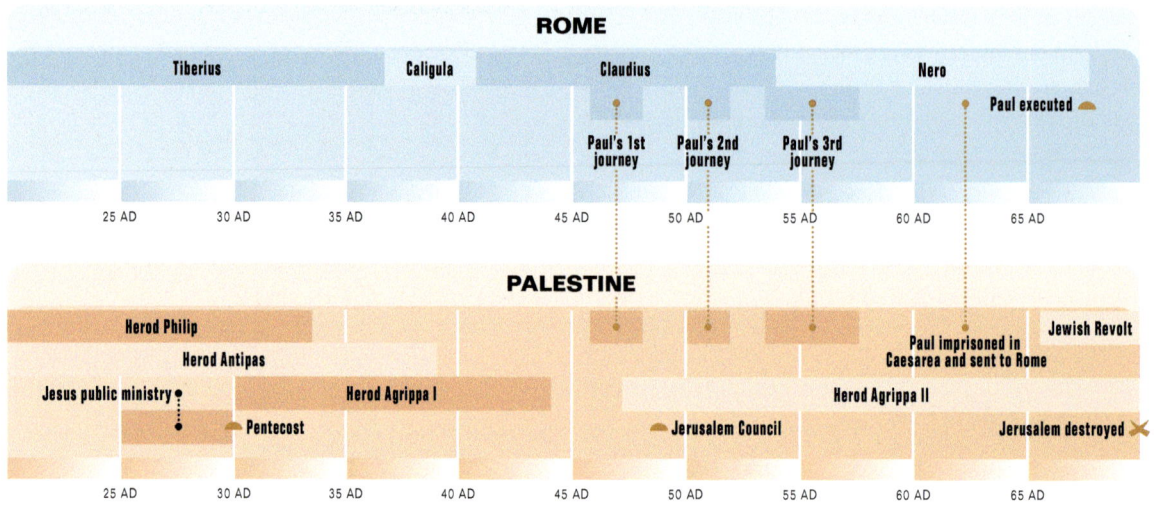

ROME

Tiberius		Caligula		Claudius		Nero	

Paul executed

Paul's 1st journey   Paul's 2nd journey   Paul's 3rd journey

25 AD   30 AD   35 AD   40 AD   45 AD   50 AD   55 AD   60 AD   65 AD

PALESTINE

Jewish Revolt

Herod Philip

Herod Antipas

Paul imprisoned in Caesarea and sent to Rome

Jesus public ministry   Herod Agrippa I   Herod Agrippa II

Pentecost   Jerusalem Council   Jerusalem destroyed

25 AD   30 AD   35 AD   40 AD   45 AD   50 AD   55 AD   60 AD   65 AD

THE EXPANSION OF THE CHURCH IN PALESTINE

Acts describes the growth of Christianity from its beginning in Jerusalem, its expansion into Judea and Samaria, and its spread throughout the Roman world (Acts 1:8). After Jesus' ascension the disciples gathered in Jerusalem. Acts 2 records the outpouring of the Holy Spirit on the disciples at the Feast of Weeks/Pentecost. At the time, Jerusalem was filled with Jewish pilgrims from all over the Roman world — a number of whom were converted to the new faith.

Opposition to the early church grew in Jerusalem as more and more Greek-speaking Jews (Hellenists) joined the ranks of Jesus' followers. As the Jerusalem church scattered throughout Judea and Samaria (8:1), the believers shared their faith with others. For example Philip traveled to a city in Samaria (8:5), where people were converted. Indeed, even Peter and John, who had come to pray for the new converts, willingly preached the gospel in Samaritan villages on their return trip to Jerusalem (8:25).

Philip also traveled south and west of Jerusalem, on the road from Bethlehem to Betogabris. There he met an Ethiopian official riding in a chariot and reading from Isaiah 53. After Philip explained the meaning of the passage, the Ethiopian believed, was baptized, and "went on his way rejoicing"

▼ *Jerusalem: paved street with shops on the west side of the Temple Mount (vertical wall on right). The tumble of boulders is from the destruction of Jerusalem in AD 70 by the Romans.*

(vv. 26 – 39), going to Gaza and then west into Africa. Philip then moved into the Philistine Plain (Azotus; 8:40) and eventually settled in Caesarea (21:8).

Peter was also active in the coastal plain area, healing Aeneas at Lydda (Acts 9:32 – 35) and raising Tabitha from the dead (9:36 – 42) at Joppa. From there he accepted an invitation to go to the house of Cornelius, a centurion living in Caesarea. Cornelius and others believed. Thus it was from Caesarea that the gospel began to make inroads into the Gentile world.

Meanwhile, the persecution of the church continued in Jerusalem and Judea. Saul, a zealous Phari- see armed with official sanction, traveled to Damascus in order to persecute the believers, but the risen Lord appeared to him and he became a believer (see next chapter).

When Jesus died, Pilate was the Roman governor of Idumea, Judea, and Samaria, while Herod the Great's sons Antipas and Philip still held their positions in the north. Philip died in AD 34, and his territory was trans- ferred to the province of Syria. But in AD 37 the districts were detached from Syria when the Roman emperor Caligula appointed Herod Agrippa I (AD 37 – 44) as the ruler over Phil- ip's old domain. When in AD 39 Antipas foolishly requested an improvement of his position, he was banished to Gaul, and Galilee and Perea were added to the realm of Herod Agrippa I.

In AD 41 the emperor Claudius, grateful for Agrippa's assistance in helping him secure the throne, added Samaria, Judea, and Idumea to his holdings. With these additions, Agrippa's kingdom was as extensive as that of his grandfather,

The Journeys of the Apostles:
→ Philip
→ Peter
→ Paul

▼ *Gamala: view to west-southwest of the city, where 9,000 Jews died attempting to defend the city against the Romans — AD 67.*

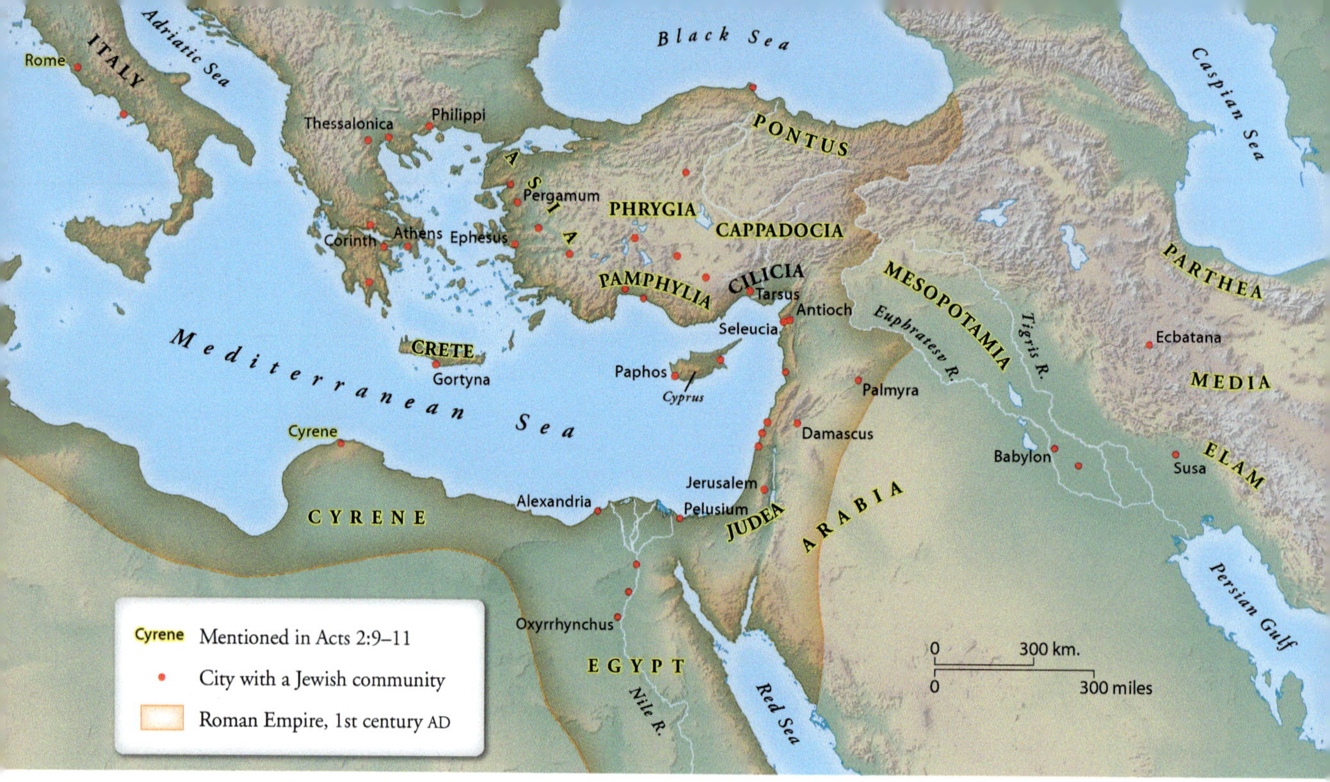

Black Sea
Caspian Sea
Adriatic Sea
Rome
ITALY
Thessalonica
Philippi
PONTUS
ASIA
Pergamum
PHRYGIA
CAPPADOCIA
Corinth
Athens
Ephesus
PARTHEA
PAMPHYLIA
CILICIA
Tarsus
MESOPOTAMIA
Antioch
Seleucia
Euphrates R.
Tigris R.
Ecbatana
Mediterranean Sea
CRETE
Gortyna
Paphos
Cyprus
Palmyra
MEDIA
Damascus
Cyrene
Babylon
ELAM
Susa
Jerusalem
Alexandria
Pelusium
JUDEA
ARABIA
CYRENE
Persian Gulf
Oxyrrhynchus
EGYPT
Nile R.
Red Sea
0
300 km.
0
300 miles
Cyrene Mentioned in Acts 2:9–11
City with a Jewish community
Roman Empire, 1st century AD

Herod the Great. However, at the peak of his power Agrippa was struck down with a terminal illness, and he died in Caesarea in AD 44 (Acts 12:19 – 23; Josephus, *Ant.* 19.8.2 [343 – 52]).

After Agrippa I's death at Caesarea in AD 44, inept and offensive procurators ruled much of Palestine. Various Jewish groups attempted to revolt, but none was successful. In the meantime, Herod Agrippa II was granted more and more territory by the Romans, so that by the time of the Jewish revolt (AD 66 – 70), he was in control of Gaulanitis, Batanea, Auranitis, Trachonitis, and portions of Galilee.

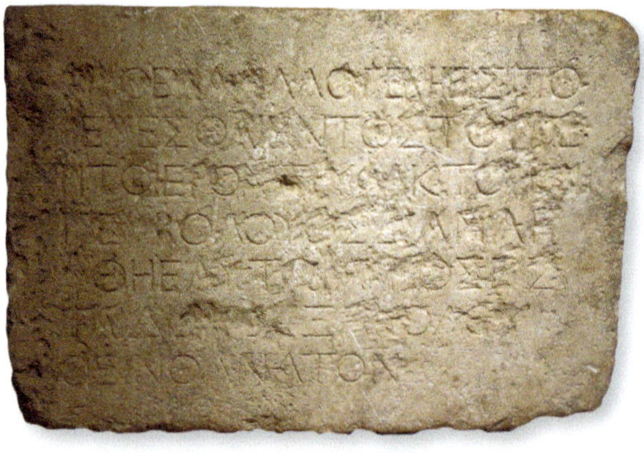

Greek inscription, found in Jerusalem, forbidding Gentiles to enter into the more sacred precincts of the temple — Paul was accused of violating this ban (Acts 21:27 – 29)

During this time Paul was traveling on his three great missionary journeys and returning to Judea after each journey. After his third journey he was imprisoned in Jerusalem, having been accused of bringing a Gentile into the temple area. Because of a plot on his life, he was transferred by night to Caesarea via Antipatris. Paul spent over two years imprisoned at Caesarea, during which time he appeared before two different procurators, Felix and Festus, as well as before the ascending Jewish king, Agrippa II. In the end, Paul appealed to Caesar.

During the tenure of the Roman governor Florus (AD 64 – 66), Jews

▲ *Rome: Arch of Titus at the entrance to the Forum, depicting the menorah being displayed in Rome after the Jerusalem temple had been destroyed in AD 70.*

all over Palestine revolted. In Jerusalem they seized the Temple Mount and the Antonia Fortress, and by the end of the summer of AD 67 all of Jerusalem was under Jewish control. In response, the Roman legate of Syria marched south with the Twelfth Legion, but he failed to retake Jerusalem.

Although the Jewish rebel forces faced serious internal divisions, a Jewish government was established and military commands were set up. The emperor Nero sent his general Vespasian to crush the revolt. Vespasian established his headquarters in Ptolemais. His first objective was to secure the northern part of the country. After retaking Sepphoris, he laid siege to the fortress of Jotapata. Although most of the

AGRIPPA II'S KINGDOM UNTIL THE JEWISH REVOLT

Legend:
- Held by Agrippa II AD 48–53
- Added to Agrippa II in AD 53
- Added to Agrippa II in AD 54
- Area under Roman procuratorial rule
- Agrippa II's kingdom in AD 61

London, British Museum

▲ *Head of Titus: as general, he destroyed the temple and Jerusalem and later (AD 79–81) ruled as emperor of Rome.*

the spring of AD 68 fighting resumed, and Vespasian's goal was first to isolate and then to capture Jerusalem. In June Emperor Nero committed suicide, but Vespasian was able to keep up the pressure in Judea. By midsummer of AD 69 only Jerusalem, the Judean Desert, Masada, and Machaerus remained in Jewish hands.

In the summer of AD 69 Vespasian's troops declared him emperor, and in the summer of AD 70 his son Titus captured Jerusalem. The Romans took the Temple Mount and set fire to the temple. The upper (western) portion of Jerusalem held out for a few weeks longer, but it too fell to the Romans. Although Jerusalem had been captured and destroyed, it was not abandoned, and the Romans stationed the Tenth Legion there to prevent further insurrection.

The capture of Jerusalem in AD 70 marks the end of the Jewish revolt, although the Romans still had to take Masada on the shore of the Dead Sea. There, 960 of the 967 Jewish defenders decided to commit suicide.

In the early second century (AD 132–35) the Jewish people staged a second revolt. Simeon Bar Kokhba was declared the leader of the Jewish people by the highly respected Rabbi Akiba. After some initial success by the Jews, the Romans sent several legions to suppress the revolt. Bar Kokhba abandoned Jerusalem and retreated to Beththter, 7 miles southwest of Jerusalem. The Romans laid siege to his fortress, and he and his garrison were annihilated. Emperor Hadrian ordered Jerusalem to be destroyed and rebuilt as a Roman colony named Aelia Capitolina, and Jews were forbidden to enter the city. He changed the name of the province from Judea to Palestine.

defending garrison died, their commander, Josephus, saved his life by surrendering to the Romans. Vespasian then regained the area around the Sea of Galilee. By the end of AD 67 all of Galilee was under Roman control.

Roman troops marched south along the coast, capturing Joppa, Jamnia, and Azotus. To the east they secured Samaritan territory in the Mount Ebal and Mount Gerizim region. In

SYRIA

Tyre

First attack of Twelfth Legion to regain Jerusalem
Attacks by Vespasian to quell the revolt
Titus's capture of Jerusalem
Area in revolt
Area partially in revolt

Cadasa

P H O E N I C I A

Gischala

Gamala

**UPPER
GALILEE**

Ptolemais

Jotapata

Taricheae

*Sea of
Galilee*

Sepphoris

Tiberias

Gaba

**LOWER
GALILEE**

Philoteria

D E C A P O L I S

Mediterranean Sea

Caesarea

Scythopolis

SAMARIA

Mt. Ebal ▲

Sebaste
(Samaria)

Jordan R.

Mt. Gerizim ▲

Acrabeta

Coreae

Gerasa

Gadora

Antipatris

P E R E A

Joppa

Lydda

Gophna

Jamnia

Beth
Horon

Jericho

Abila

Julias

Emmaus

Jerusalem

Besimoth

Azotus

Bethther

Capture of
Jerusalem AD 70

JUDEA

Judean Desert

N A B A T E A N S

Betogabris

Capharabis

Caphartobas

Hebron

*Dead
Sea*

Machaerus

Gaza

En Gedi

0   10 km.
0   10 miles

AD 73–Romans
lay siege to Masada

Masada

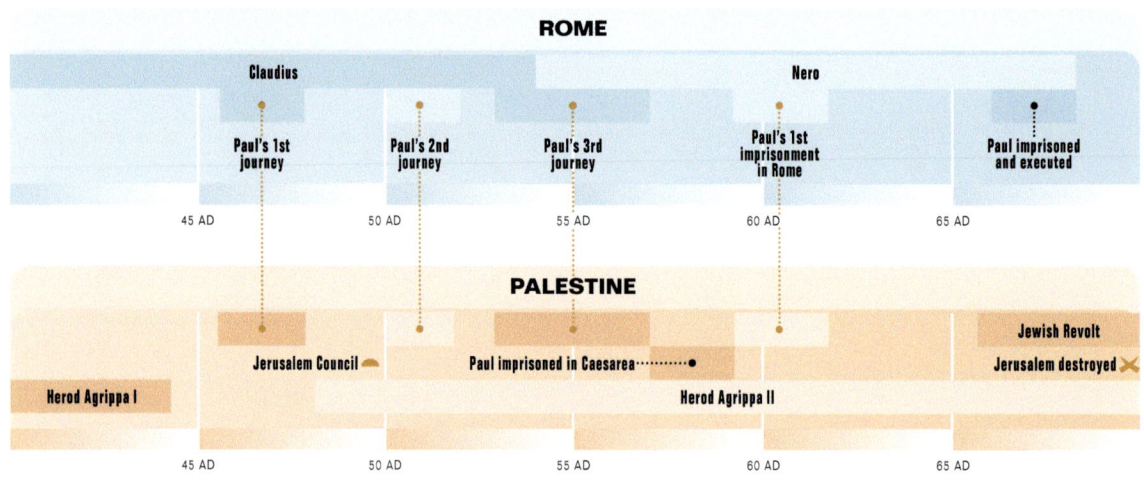

	ROME				
	Claudius			Nero	
	Paul's 1st journey	Paul's 2nd journey	Paul's 3rd journey	Paul's 1st imprisonment in Rome	Paul imprisoned and executed
	45 AD	50 AD	55 AD	60 AD	65 AD

	PALESTINE				
	Jerusalem Council	Paul imprisoned in Caesarea			Jewish Revolt / Jerusalem destroyed
Herod Agrippa I				Herod Agrippa II	
	45 AD	50 AD	55 AD	60 AD	65 AD

THE JOURNEYS OF PAUL

The Early Life of Saul

S aul (later called Paul) was born in the Greco-Roman city of Tarsus and was sent to Jerusalem to study under Rabbi Gamaliel (Acts 22:3; cf. 5:34). There, he witnessed the stoning of Stephen. But sometime later, on the road to Damascus, he himself confessed Jesus as the Messiah (Acts 9). After spending a short time in Damascus, Paul retired to Arabia (Gal 1:17). After briefly returning to Damascus, he traveled to Jerusalem for a short visit and then departed for Tarsus, his home city in Cilicia (Acts 9:26–30).

Antioch on the Orontes River was the major city of Syria. It was a leading commercial center, located at the western end of land routes leading from Mesopotamia to the Mediterranean. It was predominantly Gentile, but a considerable number of Jews lived there. Some Christians who fled Judea because of persecution after the stoning of Stephen sought refuge in Antioch. Because of their success in sharing the new faith, the Jerusalem church sent Barnabas to investigate the situation (Acts 11:22). After spending some time in Antioch, he went to

Tarsus to seek Saul's help. From about AD 43–45, Barnabas and Saul ministered together at Antioch, where believers were first called "Christians" (v. 26).

▼ *Tarsus: one of the main streets from the days of Paul. Note the basalt paving stones, the white limestone curbing, and the remains of buildings that line the right side of the street.*

Paul's First Missionary Journey (Acts 13:4 – 14:28)

The Holy Spirit then set aside Barnabas and Saul to be ministers elsewhere in the Roman world. From Seleucia, Antioch's harbor, Saul set out on the first of three missionary journeys (ca. AD 46 – 48). Barnabas and Saul — accompanied by John Mark, Barnabas's cousin — sailed to Cyprus, landing at Salamis. There they preached in the synagogue before moving overland to the administrative capital, Paphos. At Paphos, the proconsul of Cyprus, Sergius Paulus, was converted.

Next Paul and Barnabas sailed northwest to ancient Pamphylia and proceeded up the Kestros River to Perga, one of the largest cities of the province. At Perga, for unknown reasons John Mark left them and returned to Jerusalem (Acts 13:13). Paul and Barnabas headed north, up into the rugged Taurus Mountains, entering the area called Pisidia. From there they continued north into Phrygia to the city of Antioch (called "Pisidian Antioch," Acts 13:14).

Antioch was the administrative center of southern Galatia. Paul and Barnabas preached in the synagogue for several Sabbaths. Although the Jews were not overly receptive to their message, the Gentiles were, and the gospel message spread "through the whole region" (v. 49). Opposition developed, however, and Paul and Barnabas were expelled from the city and moved on to Iconium. There, Paul and Barnabas again preached in the synagogue, and a large number of Jews and Gentiles believed. But opposition and threats on their lives

PAUL'S SECOND MISSIONARY JOURNEY

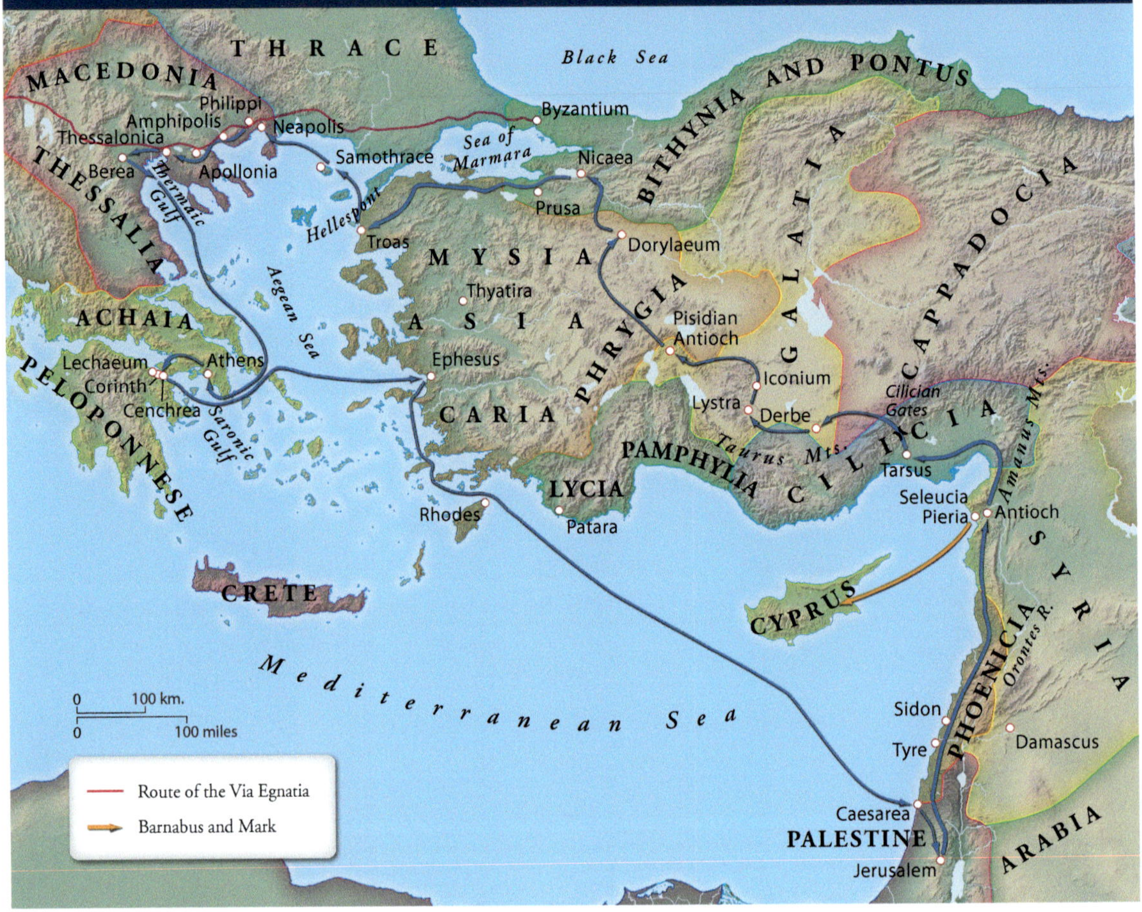
THRACE
MACEDONIA
Black Sea
BITHYNIA AND PONTUS
Philippi
Amphipolis
Thessalonica
Neapolis
Byzantium
Sea of Marmara
Samothrace
Nicaea
Berea
Apollonia
THESSALIA
Thermaic Gulf
Hellespont
Prusa
GALATIA
CAPPADOCIA
Troas
Dorylaeum
MYSIA
Thyatira
Aegean Sea
ACHAIA
ASIA
PHRYGIA
Pisidian Antioch
Lechaeum
Athens
Ephesus
Corinth
Iconium
PELOPONNESE
Cenchrea
CARIA
Lystra
Derbe
Cilician Gates
Saronic Gulf
Taurus Mts.
CILICIA
Amanus Mts.
PAMPHYLIA
Tarsus
LYCIA
Seleucia Pieria
Antioch
Rhodes
Patara
SYRIA
CRETE
CYPRUS
PHOENICIA
Orontes R.
Mediterranean Sea
0 100 km.
0 100 miles
Sidon
Damascus
Tyre
Route of the Via Egnatia
Barnabus and Mark
Caesarea
PALESTINE
ARABIA
Jerusalem

VI PAVLLI SER
LSER

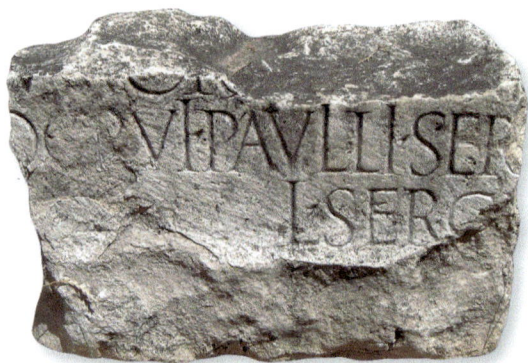

▲ *Sergius Paulus inscription from Pisidian Antioch. Possibly Sergius Paulus, whom Paul converted at Paphos on Cyprus (Acts 13), had landholdings here.*

forced them to flee, this time to the Lycaonian cities of Lystra and Derbe.

At Lystra Paul and Barnabas healed a man crippled from birth, and the Lycaonians thought that the gods were visiting them; they identified Barnabas with Zeus and Paul with Hermes. Paul and Barnabas dissuaded the townspeople from worshiping them, but when Jews from Antioch and Iconium arrived, they incited the populace to stone Paul. This lack of order and justice suggests that Roman presence was minimal there. There was no strong Jewish presence in Lystra, since no synagogue is mentioned, although Timothy, whose mother was Jewish, was from this city (Acts 16:1).

Paul and Barnabas traveled east to Derbe. After preaching in that city, they retraced their steps and strengthened the churches they had started. They went south through the Taurus Mountains and arrived at the port of Attalia and sailed back to Antioch.

Because of this ministry of Paul and Barnabas, many Gentiles began entering the church directly. This raised the question of the Gentile converts' relationship to the Mosaic law; the issue was decided at a conference in Jerusalem (Acts 15; ca. AD 49/50). Armed with this verdict, Paul and Barnabas returned to Antioch on the Orontes.

Paul's Second Missionary Journey (Acts 15:36 – 18:22)

After some time, Barnabas and John Mark sailed to Cyprus, while Paul and Silas headed toward Asia Minor. On this second journey (ca. AD 50 – 52), they went north from Antioch and traveled through the Amanus Mountains to Cilicia. Passing through Tarsus they headed through the Cilician Gates. They continued west, sharing the decision of the Jerusalem council with the churches at Derbe, Lystra, Iconium, and Pisidian Antioch. At Lystra, Timothy joined Paul and Silas.

Instead of ministering to cities in Asia, Mysia, and Bythinia, the Holy Spirit directed Paul and Barnabas to head toward Troas, a coastal city that enjoyed great prosperity as a Roman colony. There Paul evidently met Luke, a physician and the author of Luke-Acts. In response to a vision of a man from Macedonia, Paul and his party (now including Luke; see "we" in Acts 16:11) sailed to Europe.

▲ *Corinth: the temple of Apollo was over 500 years old by the time that Paul visited Corinth on his second journey.*

The ship landed in Macedonia at Neapolis, and Paul and company continued inland to Philippi, situated on the Via Egnatia — an important Roman road. As a Roman colony Philippi was populated predominantly by Gentiles, for there were not enough Jews to warrant a synagogue.

At a place of prayer used by Jewish women by the River Gangites, near the city, Paul met Lydia, the purple-cloth dealer from Thyatira. Paul ministered in Philippi while staying at her house. One day he healed a demon-possessed slave girl; as a result her owners had Paul and Silas thrown into prison. After an earthquake in the middle of the night and the subsequent conversion

▼ *Erastus inscription at Corinth: Erastus was a wealthy official in Corinth, and this inscription tells how he laid the street at his own expense (cf. Rom 16:23).*

The Seven Churches of Asia (Rev. 1–3):
addressed by John, some founded by Paul

Route of the Via Egnatia

0   50 km.
0   50 miles

of the prison guard and his family, the leaders of Philippi begged Paul and Silas to leave the city, and the two men complied.

From Philippi Paul and Silas traveled west along the Via Egnatia to Thessalonica. This city was not only a district capital but also the chief port for all of Macedonia. Paul and Silas preached in the synagogue for three Sabbaths, and a number of Jews, God-fearing Greeks, and prominent women believed. But because of strong opposition Paul and Silas left the city.

They next moved southwest to Berea, where they entered the synagogue and preached. The people of Berea were known for their desire to study the Scriptures, and a number of Jews, as well as Greek men and women, believed. However, Jews from Thessalonica agitated the crowds of Berea, so Paul departed for Athens by ship (Acts 17:14 – 15).

Athens was no longer the administrative capital of southern Greece (the Roman province of Achaia), but it was still a renowned cultural and intellectual center. Paul preached in the marketplace and synagogue and was invited by a group of philosophers to address the assembly called the Areopagus. Although a number of Athenians were converted, there does not seem to have been a ready acceptance of the gospel; so Paul left for Corinth.

Corinth was a bustling administrative and commercial center. It had been refounded as a Roman colony in 44 BC. It owed its prosperity to its geographical location — just south of the narrow isthmus that connected the Greek mainland with the Peloponnese. The ancients preferred to portage passengers and cargo across the isthmus rather than take the more dangerous trip around the Peloponnese. In addition, crowds and revenue were drawn to the city because of the biannual Panhellenic, athletic, musical, and poetic contests held at nearby Isthmia. As a "tentmaker" (Acts 18:3) Paul may have serviced the sailing ships and made tents and shelters for the visitors to the games.

The crowds eagerly received the gospel. Paul probably penned both of his letters to the Thessalonians at this time. After an eighteen-month stay Paul left Corinth via the port of Cenchrea for a sea journey to Ephesus. Stopping there for only a brief time, he continued on to Caesarea, went up to Jerusalem, and reported to the church the results of his journey. Then he returned north to his home base, Syrian Antioch.

Paul's Third Missionary Journey (Acts 18:23 – 21:14)

In AD 53, Paul set out on his third journey (ca. AD 53 – 57). He retraced the route of his second journey to Pisidian Antioch, this time continuing west to Ephesus. Ephesus had become an important commercial center, since caravan routes from the east converged there, and from there shipping lanes to the west originated. Paul spent three years ministering in Ephesus; most likely he or his converts carried the gospel message to additional cities in Asia such as those addressed in Revelation 1 – 3.

The success of the gospel in Ephesus led to a serious decline in business associated with the worship of Artemis, whose magnificent temple stood there. Thus Demetrius, a silversmith, incited the Ephesians against Paul and other Christians, but the town clerk was able to dissuade the mob from carrying out any illegal acts.

Soon afterward, Paul set out for Macedonia (Acts 20:1; 2 Cor 2:12 – 13), probably revisiting the churches at Philippi, Thessalonica, and Berea. Eventually he continued south and spent three months wintering in Corinth. During this time he likely wrote his famous letter to the church at Rome, informing them of his intention to visit them after a visit to Jerusalem.

When spring came, Paul traveled overland to Philippi and then sailed for Troas. After spending seven days there, he walked the 20 miles overland to Assos and then sailed down the coast to Miletus. From Miletus, he summoned the elders of the Ephesian church, who traveled there to spend a few days with their beloved teacher.

After a tearful goodbye, Paul and his party headed by ship for Jerusalem. After disembarking at Tyre and against the advice of some Christians, he continued on to Jerusalem, where he greeted the elders of the church and completed the purification rites associated with a vow he had made. Jerusa-

▼ *Patara: Probably the oldest preserved lighthouse in the world. Paul changed ships at Patara on his return from his third journey (Acts 21:1).*

MOESIA

THRACE

Black Sea

Rome
Three Taverns
Forum of Appius
LATIUM
Puteoli
APULIA
MACEDONIA
BITHYNIA AND PONTUS
Adriatic Sea
LUCANIA
EPIRUS
GALATIA
CAPPADOCIA
Rhegium
SICILY
Syracuse
ACHAIA
Aegean Sea
ASIA
MESOPOTAMIA
PELOPONNESE
PAMPHYLIA
CILICIA
Adramyttium
Cnidus
LYCIA
MALTA
Myra
Seleucia
Antioch
SYRIA
CRETE
CYPRUS
PHOENICIA
M e d i t e r r a n e a n   S e a
Sidon
Gulf of Syrtis
L I B Y A
Caesarea
Antipatris
Jerusalem
Alexandria
PALESTINE
ARABIA
EGYPT
Red Sea

Salmone
Phoenix
C R E T E
Cauda
Lasea
Fair
Havens

0   50 km.
0   50 miles

0   150 km.
0   150 miles

▼ *Ephesus: view toward the now-silted harbor from the theater at Ephesus (Acts 19:23–41)*

lem ended up being the terminal point of his third journey, for here he was arrested. After being transported to Caesarea, he remained imprisoned there for two years (ca. AD 57–59) before finally appealing to Caesar for justice.

Paul's Journey to Rome (Acts 27:1–28:16)

For the trip to Rome, Paul was placed in the custody of a centurion named Julius. Along with a small party that included Aristarchus and probably Luke, Paul was placed on a ship that hugged the coast of Asia Minor until Myra, a major port of call for grain ships bound for Rome. There Paul and his

▲ *Assos: The temple of Athena. Paul visited Assos on the return leg of his third journey (Acts 20:13).*

party transferred onto a cargo ship that was to sail directly to Italy. Because of adverse winds, the ship was not able to reach Cnidus on the southwestern tip of Asia Minor. Instead, it sailed south, intending to pass on the lee side of the island of Crete.

By this time it was late fall, and in their attempt to reach the desirable port of Phoenix on Crete, a "northeaster" (Acts 27:14) wind sprang up, and the ship was driven off course. As it passed on the lee side of the island of Cauda, the ship had to be lashed together. For two weeks it tossed about in the Mediterranean Sea and eventually ran aground on the island of Malta. All 276 passengers were saved, but the ship and its cargo were lost.

After wintering three months on Malta, they sailed to Italy, landing at the port of Puteoli. On the road to Rome, at the Forum of Appius and at Three Taverns, Paul was greeted by Christians from the capital. The Rome of Paul's day was a huge city with a population of about 1,000,000. As the capital of the Roman Empire, it boasted imperial palaces on the Palatine Hill, temples of Jupiter and Juno on the Capitoline Hill, and theaters, amphitheaters, hippodromes, and other monuments. But its beauty was tempered by the fact that over half its population were slaves, and many others lived in squalid conditions in high-rise apartment buildings of four and five stories and depended on the free distribution of food.

The book of Acts concludes with Paul residing in Rome for two years, under house arrest, without ever having had his case go to trial. According to tradition, Paul was released from prison around AD 62 and traveled to various parts of the Mediterranean world — probably to Crete (Titus 1:5) and possibly to Spain. Several years later, he was arrested and imprisoned again, during which time he penned his final letter (2 Timothy). Tradition claims that the Roman emperor Nero had Paul executed outside the walls of Rome.

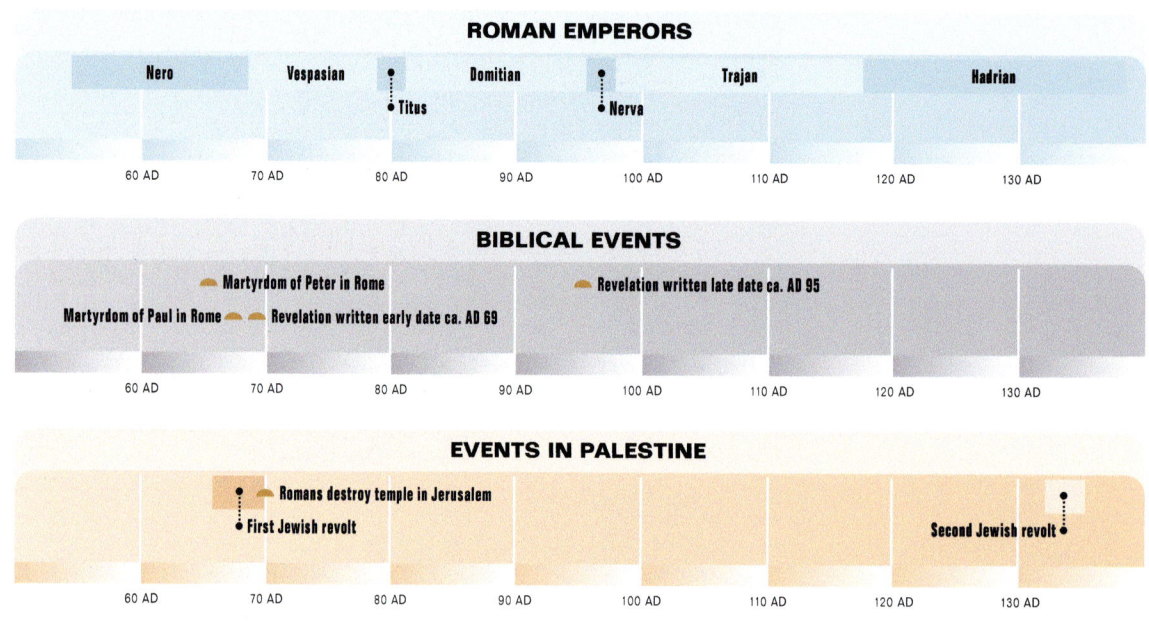

ROMAN EMPERORS

Nero	Vespasian	Titus	Domitian	Nerva	Trajan	Hadrian

60 AD   70 AD   80 AD   90 AD   100 AD   110 AD   120 AD   130 AD

BIBLICAL EVENTS

Martyrdom of Peter in Rome

Revelation written late date ca. AD 95

Martyrdom of Paul in Rome   Revelation written early date ca. AD 69

60 AD   70 AD   80 AD   90 AD   100 AD   110 AD   120 AD   130 AD

EVENTS IN PALESTINE

Romans destroy temple in Jerusalem

First Jewish revolt

Second Jewish revolt

60 AD   70 AD   80 AD   90 AD   100 AD   110 AD   120 AD   130 AD

THE SEVEN CHURCHES OF REVELATION

Early Christian traditions associate the apostle John with Ephesus, and he may have written his gospel and his three letters from here. During the Byzantine period Christianity flourished at Ephesus, and the Third Ecumenical Council was held there in AD 431.

▼ *Patmos: the barren Aegean island to which John was exiled (Rev 1:9)*

I n Revelation 1 – 3, John addresses seven churches located in the Roman province of Asia. Beginning with Ephesus, the churches are addressed in a clockwise fashion, ending at Laodicea. This may be the order that the carrier of the document followed.

Ephesus (1:11; 2:1 – 7)

Ephesus was the closest church to the desolate island of Patmos, where John may have been exiled and where he received his "revelation" (Rev 1:9 – 11). It was the capital of Asia Minor. It had a population of about 250,000 people and was an important trade center. It housed numerous temples and cults. Note how Paul's letter to the Ephesians addresses the priority of worshiping Jesus, rather than Caesar, as "Lord."

Smyrna
(1:11; 2:8 – 11)

Smyrna is 36 miles north of Ephesus. This Roman city was on a sheltered bay at the western end of the Hermus River Valley. It was the first city in Asia Minor to establish a temple dedicated to Dea Roma, the goddess Roma (ca. 195 BC). In AD 26 Tiberius granted it permission to establish a temple for worshiping the emperor.

The church there is encouraged to remain faithful in light of the coming persecution. The early church father Polycarp was martyred in the stadium of Smyrna at the age of 85 (AD 156). The "victor's crown" mentioned in Revelation 2:10 may refer to the wreath given to a victorious athlete. The modern city of Izmir is built over the ancient remains of Smyrna.

Pergamum
(1:11; 2:12 – 17)

Pergamum was a magnificent city with a population of about 100,000. The upper city rises about 900 feet above the surrounding plain and held many temples and altars. In the lower city significant remains of the Asclepieion (an ancient healing center) are still visible.

In 129 BC, when Rome created the province of Asia, Pergamum became its early capital. In 29 BC it was the first city to be granted permission to establish a

▲ *Pergamum: the great theater and beyond it, by the tree, the altar of Zeus (the throne of Satan? Rev 2:13)*

THE SEVEN CHURCHES OF REVELATION

Black Sea

Byzantium

Sea of Marmara

Alexandria Troas

Assos

Pergamum

A S I A

Thyatira

Sardis

Smyrna

Philadelphia

Hierapolis

Ephesus

Laodicea   Colosse

Patmos

Attalia

Rhodes

0   50 km.

0   50 miles

Rhodes

◆ Seven Churches of Revelation 1–3

▲ Sardis: gymnasium and palestra of this once very rich city
▼ Laodicea: calcified siphon that once brought water into the city (Rev 3:15 – 16)

Thyatira (1:11; 2:18 – 29)

Thyatira (modern Akhisar) is located 42 miles inland from the Aegean Sea. It was a major commercial center. Near the center of Akhisar, visible archaeological remains are located in a fenced-off rectangular city block. From inscriptions it is evident that guilds of bakers, bronze smiths, wool workers, potters, linen weavers, and tanners were active in the city. Such guilds often held banquets that included immoral sexual acts (cf. Rev 2:20 – 24). Lydia, converted by Paul in Philippi, originated from Thyatira (Acts 16:11-15).

Sardis (1:11; 3:1 – 6)

Sardis, in the Hermus Valley east of Smyrna, was for many years the chief city of the powerful kingdom of Lydia. It derived its wealth from gold mining, trade, and the manufacture of textiles. In 546 BC Cyrus the Persian conquered the city. The Persians built the "Royal Road" that connected Sardis with Susa, one of the Persian capitals (map on p. 93).

In the first century AD the glory days of Sardis were over. When John admonishes the church to "wake up" (Rev 3:2) and asserts that Jesus will "come like a thief" in the night (3:4), he may be alluding to the two occasions when the enemies of Sardis were able to capture its citadel because of the laxness of its defenders. The reference to true believers being "dressed in white" garments (3:4 – 5) may allude to the famous textile industry of Sardis.

Philadelphia (1:11; 3:7 – 13)

Philadelphia (modern Alashehir) was founded in the third century BC by one of the Pergamenian kings and named after Attalus II, who maintained "loyalty/love" for his brother — thus Philadelphia (lit., "brotherly love"). It housed a number of temples; in AD 17 it was destroyed by an earthquake. In Revelation 3:12 the believer who "is victorious" is compared to a pillar (stability) in the temple of God.

Laodicea (1:11; 3:14 – 27)

Laodicea was founded in the third century BC; by the first century AD it had replaced Colossae and Hierapolis as the

temple dedicated to a Roman emperor (Augustus). In Revelation 2:13 it is said that "Satan has his throne" there.

John also writes that Antipas was martyred there (2:13), but note that it is King Jesus who wields the real "double-edged sword" (2:12), not Caesar or his representatives. On the negative side, John warns the church that some must give up eating "food sacrificed to idols" and living immorally (2:14 – 16).

▲ *Laodicea: excavated street of this proud, self-satisfied city (Rev 3:14 – 22)*

chief city of the region. It was famous for its textiles, for an eye salve produced there, and for its nearby medical center. Laodicea was so wealthy that after the devastating earthquake of AD 60, it paid for its own rebuilding and declined assistance offered.

Epaphras (Col 1:7; 2:1; 4:12; Philem 23) may have brought the gospel to Laodicea, ministering also to Colossae and Hierapolis. Paul wrote a letter not only to the church at Colossae (Col 1:1 – 2), but also to the church at Laodicea (4:16).

The church at Laodicea is the only one of the seven churches not to receive any commendation. The church seemed self-satisfied (Rev 3:17). In 3:18 gold, textiles, and eye salve are all mentioned — the very commodities Laodicea took pride in. The reference to the church being lukewarm (3:15 – 16) may allude to the desirable — for medicinal purposes — hot springs at Hierapolis and to the cool stream waters running by Colossae. Laodicea had built an aqueduct to bring water to it from springs 8 miles to the southeast.

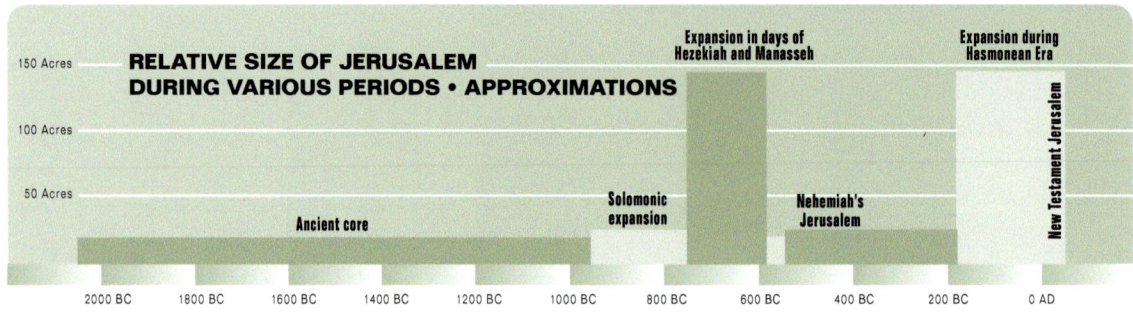

RELATIVE SIZE OF JERUSALEM DURING VARIOUS PERIODS • APPROXIMATIONS

150 Acres

100 Acres

50 Acres

Expansion in days of Hezekiah and Manasseh

Expansion during Hasmonean Era

New Testament Jerusalem

Ancient core

Solomonic expansion

Nehemiah's Jerusalem

2000 BC · 1800 BC · 1600 BC · 1400 BC · 1200 BC · 1000 BC · 800 BC · 600 BC · 400 BC · 200 BC · 0 AD

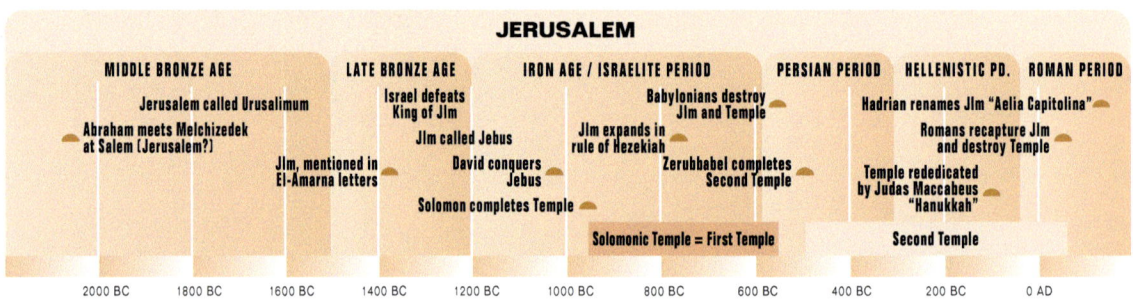

JERUSALEM

MIDDLE BRONZE AGE	LATE BRONZE AGE	IRON AGE / ISRAELITE PERIOD	PERSIAN PERIOD	HELLENISTIC PD.	ROMAN PERIOD

Jerusalem called Urusalimum

Abraham meets Melchizedek at Salem [Jerusalem?]

Jlm, mentioned in El-Amarna letters

Israel defeats King of Jlm

Jlm called Jebus

David conquers Jebus

Solomon completes Temple

Jlm expands in rule of Hezekiah

Babylonians destroy Jlm and Temple

Zerubbabel completes Second Temple

Hadrian renames Jlm "Aelia Capitolina"

Romans recapture Jlm and destroy Temple

Temple rededicated by Judas Maccabeus "Hanukkah"

Solomonic Temple = First Temple

Second Temple

2000 BC · 1800 BC · 1600 BC · 1400 BC · 1200 BC · 1000 BC · 800 BC · 600 BC · 400 BC · 200 BC · 0 AD

JERUSALEM

Jerusalem holds a special place in the hearts and minds of Jews, Christians, and Muslims. It is mentioned 667 times in the Old Testament and 139 times in the New. Although today the city boasts a population of over 770,000 people, its origins were humble.

Geography of Jerusalem

Jerusalem was located in the Hill Country of Judah, far removed from the Coastal and the Transjordanian highways. The only route that passed by it was the north–south Ridge Route, and even that ran about half a mile west of the city. A west–east road that connected Gezer on the coastal plain with Jericho in the Jordan Valley passed 5.5 miles to the north.

Jerusalem's location in the hill country, at an elevation of 2,500 feet, gave it the benefit of many natural defenses. It has a rugged and treacherous landscape that protected access to the city from the east and west. It was somewhat easier to

approach Jerusalem from the north or south, along the Ridge Route, but access to the Ridge Route was difficult.

While arid land lies to the east and south, Jerusalem itself receives ample supplies of winter rain (approximately 25 in.

▼ *Jerusalem: view toward the ancient core from the south. The golden-colored Dome of the Rock stands where the ancient temple once did. Note how the hills surround Jerusalem (Ps 121:1).*

per year), as do the hills to the west, so that a variety of crops can be grown on the hillside terraces to the north, west, and south of the city.

Biblical Jerusalem was built on two parallel north – south ridges. The western ridge, the higher and broader one, is bounded on the west and south by the Hinnom Valley (map p. 142). The narrower and lower eastern ridge is bounded on the east by the Kidron Valley, which in the Jerusalem area flows basically north to south. Both the Hinnom and the Kidron are noted in the Bible, but the valley between them, which separates the eastern and western ridges, is not. It is often called the Central or the Tyropoeon ("Cheesemakers") Valley (Josephus *War* 5.4.1 [140]).

On the north, both ridges continue to rise as they veer to the northwest. Because of the easier approaches from the north and the northwest, invading armies generally assaulted Jerusalem from a northerly direction.

Early History of Jerusalem

The earliest settlement in Jerusalem began on the 15-acre southern portion of the eastern ridge, "the old ancient core," because the only good-sized spring — the Gihon Spring (see p. 69) — was located there. Tombs from the MB I period have been found in the Jerusalem area, but there is no evidence of a settlement. During the MB II period (2000 – 1550 BC) Jerusalem is mentioned several times in the Egyptian Execration Texts as Urusalimum (meaning "foundation of the god Shalim" or "city of peace"). Although excavated building remains are few, significant portions of a thick city wall have been uncovered. This wall was apparently built about 1800 BC and continued in use, with rebuilds, until the end of the Judean monarchy (586 BC). The city remained 15 acres until it began to expand northward during the days of David and Solomon.

Two events in the life of Abraham place him in close proximity to Jerusalem. Melchizedek, the king of Salem (Gen 14:18; cf. Ps 76:2), met Abram after his rescue of Lot. Later Abraham took his son Isaac to one

▲ *Looking north up the Kidron Valley. The slopes of the City of David are on the left (west), and a corner of the Temple Mount is visible.*

THE SURROUNDINGS OF JERUSALEM

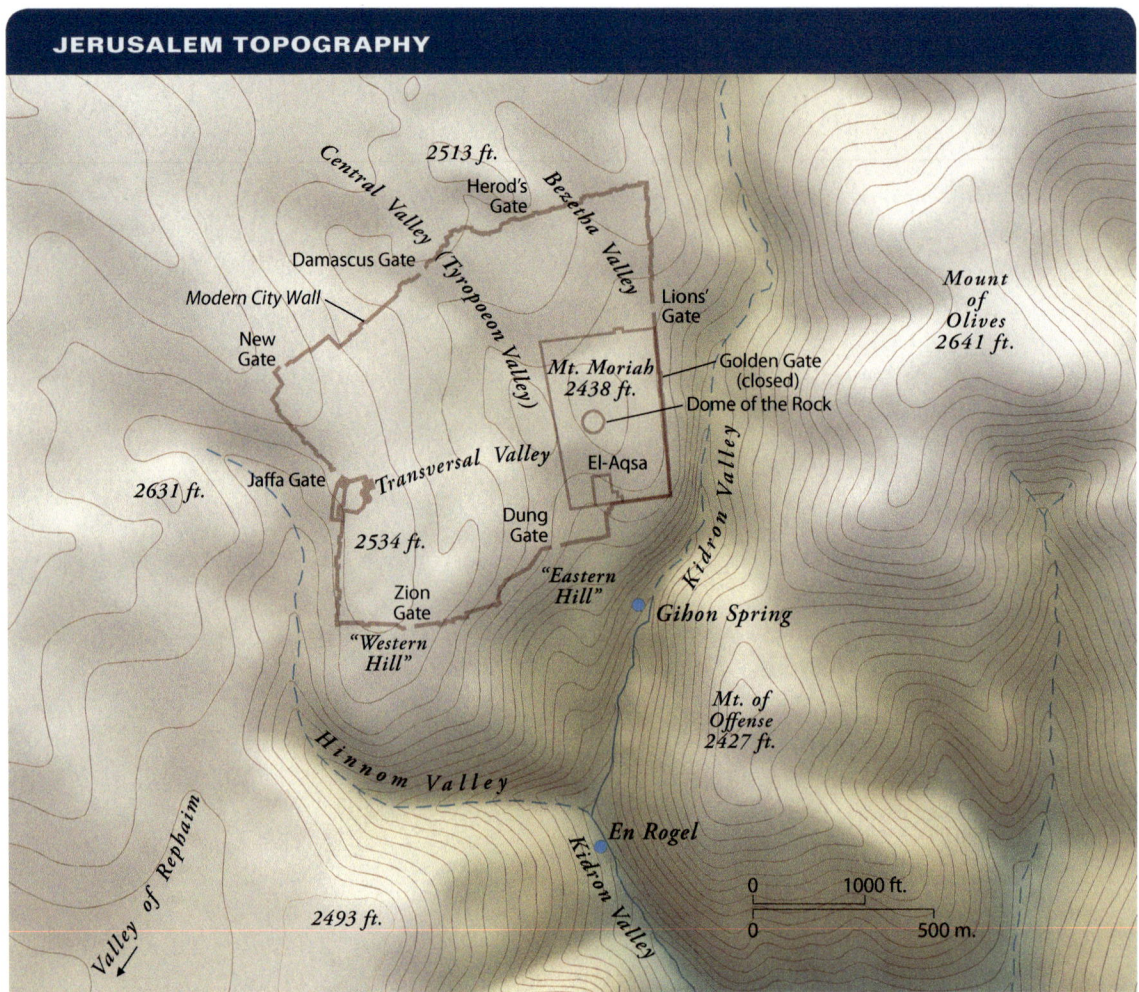

of the mountains in the "region of Moriah" to sacrifice him (Gen 22:2), the same place where Solomon built the temple (2 Chron 3:1).

Jerusalem next appears in the stories of the conquest under Joshua. When the king of Jerusalem, Adoni-Zedek, heard of the Gibeonites' treaty with Joshua, he realized that his major line of communication with the coast, and hence with Egypt, was in jeopardy. He assembled a coalition of four other Amorite kings and attacked Gibeon (map on p. 50) but was defeated by Joshua.

During the period of the judges Jerusalem came under the control of the Jebusites and was named Jebus (see Judg 19:11 – 12; cf. also Josh 15:8; 18:16). It was the Judahite David

who captured the city in his seventh year (2 Sam 5) and made it his capital. His general Joab used the *sinnor* ("water shaft," 2 Sam 5:8) to do so — an underground, rock-cut diagonal tunnel (see picture on p. 87) that led from inside the city to a large pool fed by the Gihon Spring.

The City of David

Because of Jerusalem's neutral location, it was a capital acceptable to both David's own tribe of Judah as well as to the tribes of the north. The city became David's and his descendants' personal property (called "the City of David") and the royal seat of the Davidic dynasty. David brought the ark from Kiri-

▲ *Looking south at the Dome of the Rock. Many believe this is where the ancient Israelite temple stood: the Most Holy Place under the dome, and the sacrificial altar near the smaller structure to the left (east) of the dome. There are close to 50 cisterns on the Temple Mount.*

ath Jearim to Jerusalem, which he established as the major worship center for all Israel (2 Sam 6:1 – 23; 1 Chron 13:1 – 14). David built his own palace there (2 Sam 5:11) and toward the end of his reign purchased the threshing floor of Araunah the Jebusite, a site north of and higher than the ancient city core, where Solomon eventually built the temple (2 Sam 24:18 – 25; 1 Chron 21:18 – 26).

In the fourth year of Solomon's reign (966 BC), he began building the temple, a task that took seven years. The building itself was composed of two rooms: the Holy Place, in which the ten lampstands, the table for the bread of the Presence, and the incense altar were placed; and the most sacred place, called the Most Holy Place or the Holy of Holies, in which the ark of the covenant was kept. The whole building was surrounded by courtyards in which were located the sacrificial altar, lavers, and the like. The exact location of the temple is not known, although many researchers place it in the immediate vicinity of the existing Muslim shrine called the Dome of the Rock.

To the south of the temple, but north of the ancient core of Jerusalem, Solomon built his own palace and the Palace of the Forest of Lebanon (1 Kings 7:1 – 12). It is possible that this royal acropolis was, in early times, called the Millo (NIV "the terraces"; 1 Kings 9:15, 24; 11:27) but later came to be known as the Ophel (the acropolis). Solomon strengthened the wall of Jerusalem and included the Millo/Ophel, as well as the temple area, within the confines of the wall. Thus the walled city expanded from 15 acres to about 37 acres (map on p. 144).

During the divided monarchy (930 – 722 BC), Jerusalem was attacked several times: once by the Egyptian pharaoh Shishak (925 BC; 1 Kings 14:22 – 28; 2 Chron 12:2 – 4) and once by Hazael of Aram Damascus (ca. 813 BC; 2 Kings 12:17 – 18; 2 Chron 24:17 – 24). In each instance, lavish gifts, taken from the temple treasury, bought off the aggressors.

In the days of Amaziah of Judah, however, Jehoash of Israel attacked the city and "broke down the wall of Jerusalem from the Ephraim Gate to the Corner Gate"(ca. 790 BC; 2 Chron 25:23). It is difficult, however, to pinpoint the location of these gates in the city walls.

During the eighth century BC "Uzziah built towers in Jerusalem at the Corner Gate, at the Valley Gate and at the angle of the wall" (2 Chron 26:9) as he strengthened the defenses of the city. Also during his reign (792 – 740 BC) and after, Jerusalem

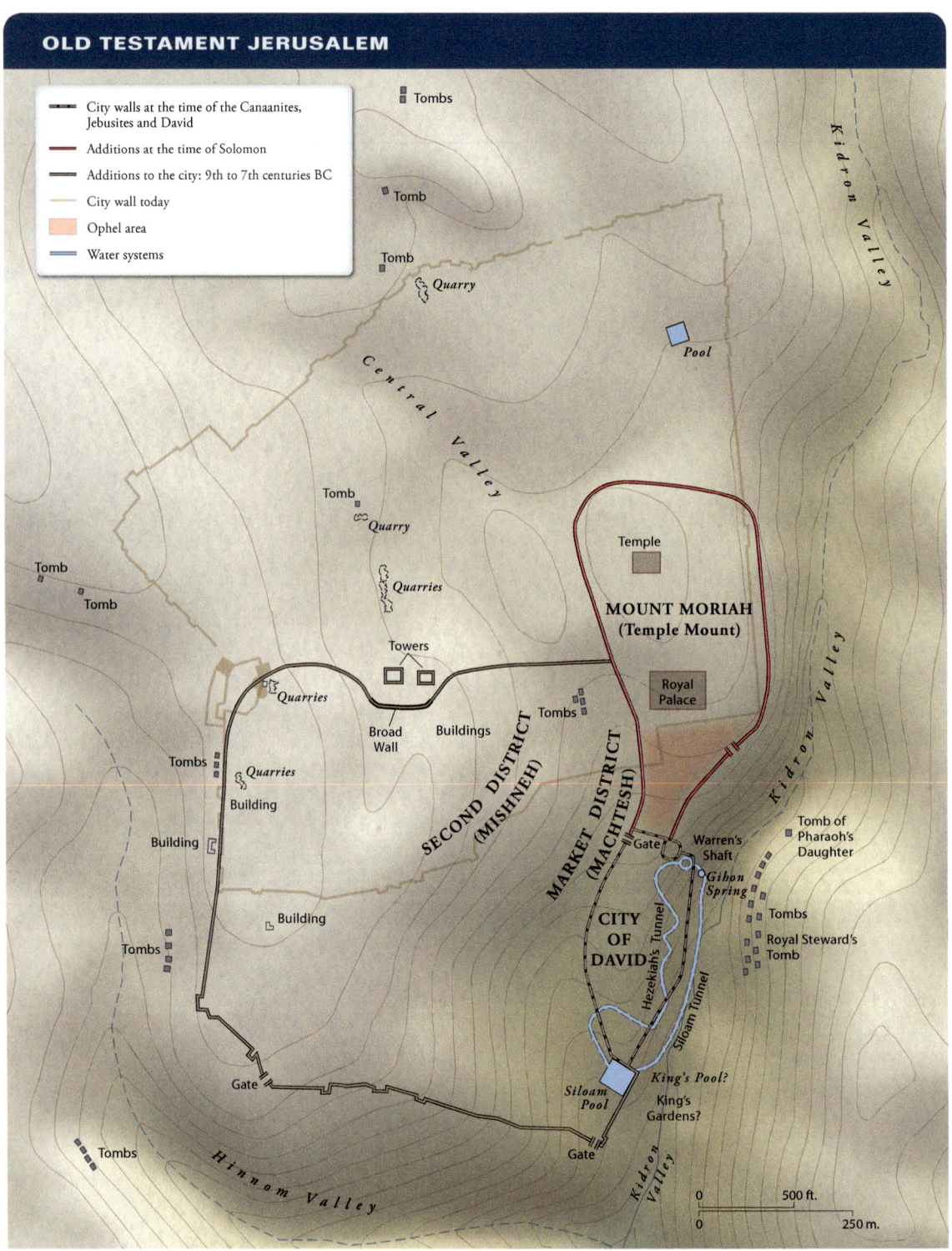

Legend:
- City walls at the time of the Canaanites, Jebusites and David
- Additions at the time of Solomon
- Additions to the city: 9th to 7th centuries BC
- City wall today
- Ophel area
- Water systems

Tombs

Tomb

Tomb

Quarry

Pool

Central Valley

Kidron Valley

Tomb

Quarry

Quarries

Temple

MOUNT MORIAH
(Temple Mount)

Tomb

Tomb

Towers

Royal
Palace

Tombs

Broad
Wall

Buildings

Quarries

SECOND DISTRICT
(MISHNEH)

MARKET DISTRICT
(MACHTESH)

Tombs

Quarries

Gate

Warren's
Shaft

Tomb of
Pharaoh's
Daughter

Building

Gihon
Spring

Building

CITY
OF
DAVID

Hezekiah's Tunnel

Tombs

Royal Steward's
Tomb

Building

Siloam Tunnel

Kidron Valley

Tombs

King's Pool?

Siloam
Pool

King's
Gardens?

Gate

Kidron Valley

Gate

Tombs

Hinnom Valley

| 0 | | 500 ft. |
| 0 | | 250 m. |

expanded westward so as to include the southern portion of the western ridge — probably because settlers from the northern kingdom moved south to avoid the Assyrian onslaught (see p. 84); they may have thought that Jerusalem would never be taken because the temple of Yahweh was there (Ps 132:13 – 18).

In the excavations in the modern Jewish Quarter of the Old City of Jerusalem, a 210-foot segment of a massive wall, 23 feet thick and in places preserved to a height of 10 feet, was discovered (see photo on p. 85). This was likely built in Hezekiah's day because of the threat of Assyrian assault (see p. 86 for the history). He enclosed the whole southern portion of the western ridge so that the total area of the walled city swelled to 150 acres and boasted a population of about 25,000.

Since the Gihon Spring was at some distance from the newly enclosed western suburb, Hezekiah devised a plan to divert the water to a spot inside the city walls, closer to the western hill. He did this by digging an underground tunnel (see photo, p. 87) that followed a serpentine path to a point in the Central Valley, which was inside the newly constructed city wall. This diversion is mentioned not only in the Bible (2 Kings 20:20), but also in a Hebrew inscription discovered at the southern end of the 1,750-foot tunnel.

Postexilic Jerusalem

But because of the continuing sins of the people and their leaders, God's judgment fell on Jerusalem in 605, in 597, and climactically in 586 BC — the year when Nebuchadnezzar destroyed both the city and the temple (see p. 89). Almost fifty years later, a large-scale return to Jerusalem began in response to the decree issued by Cyrus (539 BC). Led by Sheshbazzar, 49,897 people returned to Jerusalem from Babylon, rebuilt the temple altar, and reinstituted sacrificial worship. Not until the days of the Persian Darius, however, were Jews, led by Zerubbabel, able to actually rebuild the temple (520 – 516 BC; Ezra 6).

The second return from Babylon was led by Ezra the scribe (458 BC) and was noted for its spiritual accomplishments. The actual rebuilding of the walls took place in the days of Nehemiah (445 BC; see Neh 1 – 4; 6; 12:27 – 47). From that time until the beginning of the second century BC, not much is known about Jerusalem.

Early in the second century the Seleucid king Antiochus III defeated the Ptolemies (198 BC), and the change in rule was welcomed by most of the Jewish population. With Antiochus's support, repairs were made to the temple, and a large pool — possibly the Pool of Bethesda — was constructed (Sir 50:1 – 3).

During the reign of Antiochus IV (175 – 164 BC), however, the king and his Jewish supporters pressed for a Hellenizing program among all of the Jews. The temple in Jerusalem was desecrated and a statue of Olympian Zeus was set up in its precincts (168 BC). Other Greek structures were erected in Jerusalem, including a gymnasium and a citadel. The citadel (called the "Akra" in Greek) was built on the eastern ridge just south of the temple area and was so tall that it towered over the temple area. Although Judas Maccabeus's forces were able to retake Jerusalem, purify the temple (164 BC),

JERUSALEM AT THE TIME OF NEHEMIAH

Kidron Valley

Modern Wall

Tower of Hananel

Mount of Olives

Fish Gate

Sheep Gate

Muster Gate

Temple

East Gate

Horse Gate

Modern Wall

Valley Gate

Gihon Spring

Hezekiah's Tunnel

Stairs from City of David

Siloam Pool

Water Gate

Hinnom Valley

0   1000 ft.

0   500 m.

Modern wall

Gate

City of Nehemiah

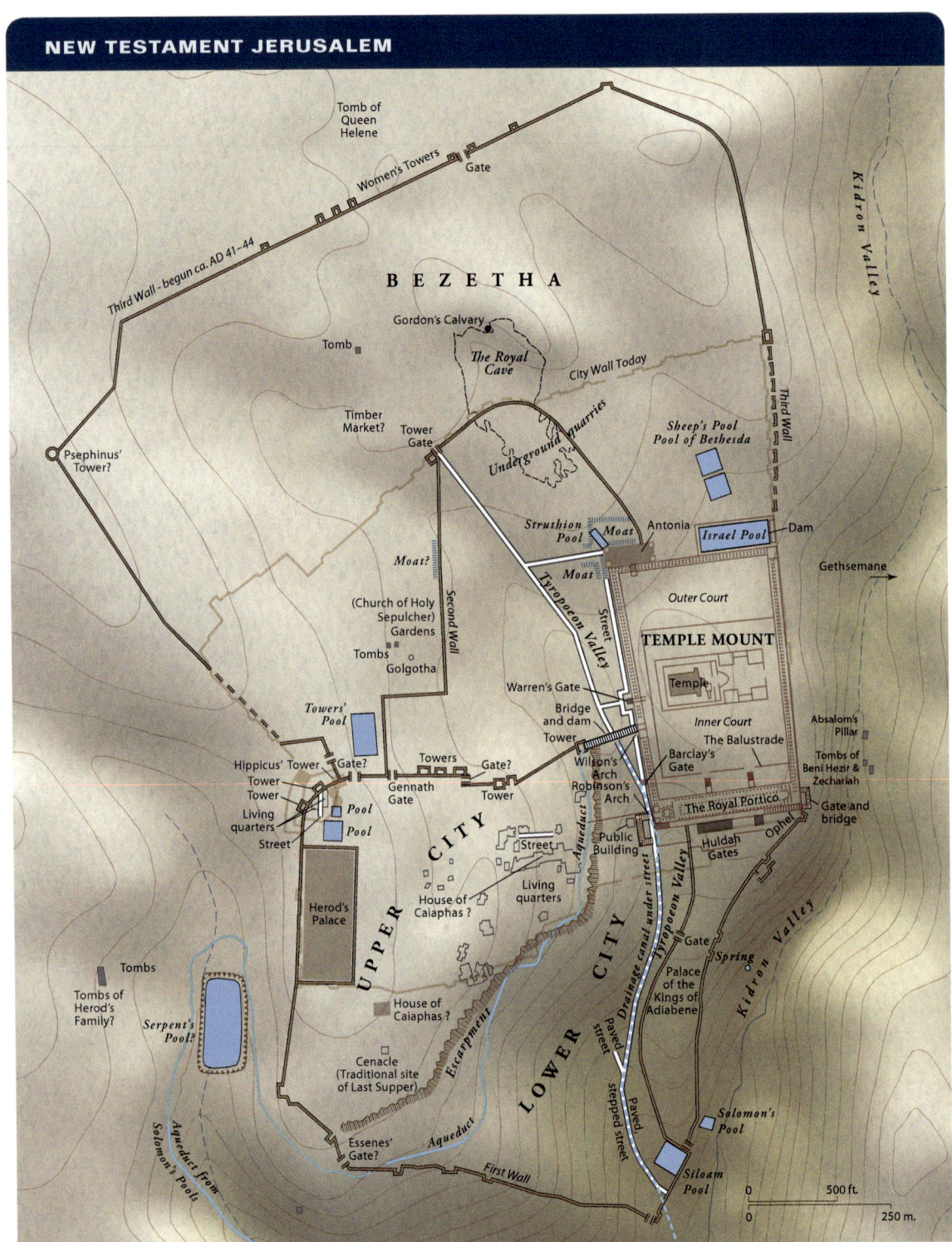

Tomb of Queen Helene

Women's Towers

Gate

Third Wall – begun ca. AD 41–44

B E Z E T H A

Tomb

Gordon's Calvary

The Royal Cave

City Wall Today

Psephinus' Tower?

Timber Market?

Tower Gate

Underground Quarries

Sheep's Pool Pool of Bethesda

Third Wall

Struthion Pool

Moat

Antonia

Israel Pool

Dam

Gethsemane

Kidron Valley

Moat?

Moat

Second Wall

Tyropoeon Valley

Street

Outer Court

TEMPLE MOUNT

Temple

(Church of Holy Sepulcher) Gardens

Tombs

Golgotha

Warren's Gate

Bridge and dam Tower

Inner Court

The Balustrade

Absalom's Pillar

Towers' Pool

Towers

Gate?

Wilson's Arch

Robinson's Arch

Barclay's Gate

Tombs of Beni Hezir & Zechariah

Hippicus' Tower

Gate?

Gennath Gate

Tower

The Royal Portico

Gate and bridge

Tower Tower

Living quarters

Pool

Pool

Street

Aqueduct

Public Building

Huldah Gates

Ophel

Street

U P P E R   C I T Y

House of Caiaphas ?

Living quarters

Drainage canal under street

Tyropoeon Valley

Gate

Spring

Kidron Valley

Herod's Palace

L O W E R   C I T Y

Palace of the Kings of Adiabene

Tombs

Tombs of Herod's Family?

House of Caiaphas ?

Escarpment

Paved street

Serpent's Pool?

Cenacle (Traditional site of Last Supper)

Paved stepped street

Solomon's Pool

Aqueduct from Solomon's pools

Essenes' Gate

Aqueduct

First Wall

Siloam Pool

0   500 ft.

0   250 m.

▲ *Looking east at the northeast corner of the recently discovered "Pool of Siloam" (John 9). Note the series of steps and platforms that lead down to the pool from left to right.*

and reestablish sacrificial worship, the Seleucid garrison in the Akra remained a thorn in the side of the Jews until Judas's brother Simon (142–135 BC) captured and demolished it (Josephus, *Ant.* 13.6.7 [215]). Simon also completed building the walls of Jerusalem, a project his brother Jonathan had begun (1 Macc 10:10–11; 13:10). (On the history of Jerusalem during the Hasmonean era, see pp. 103–7.)

Jerusalem in Roman Times

At the beginning of the period of Roman rule Jerusalem experienced great expansion, construction, and beautification under the leadership of the Roman client king, Herod the Great (37–4 BC). Pride of place must certainly go to Herod's refurbishing of the temple and the Temple Mount (see p. 112), a project that took ten years, though crews were still working

on it during Jesus' lifetime (John 2:20, ca. AD 28). Herod especially expanded the courts surrounding the temple. He doubled the size of the platform area so that it reached its present size of 36 acres. The area is now occupied by Muslim structures and is called the Haram esh-Sharif — the Noble Sanctuary. To the northwest of the temple Herod built the Antonia Fortress, which towered over the temple area and housed a garrison to monitor and control the crowds.

On the western ridge Herod built a magnificent palace for himself (see p. 112). In addition, Herod built a second wall that began near these towers — by the Gennath Gate — and ran to the Antonia Fortress, enclosing the northern "Second Quarter" of the city (Josephus, *War* 5.4.2 [146]).

The Jerusalem Jesus knew was basically the same as Herodian Jerusalem. On one of his visits to the city, Jesus healed a paralyzed invalid at the Pool of Bethesda, north of the Temple

▲ *Garden Tomb located about 260 yards north of the present-day wall of the Old City. This tradition dates back to the 19th century AD.*

Mount near the Sheep Gate (John 5:1 – 14). Portions of a double pool that could have been surrounded by "five covered colonnades" — one on each side and one in the middle separating the two pools — have been discovered just north of the Temple Mount. On another occasion Jesus healed a blind man whom he sent to the Pool of Siloam to wash (John 9).

Most of the information about Jesus in Jerusalem comes from the last week of his earthly ministry. Jesus evidently spent his nights with his friends in Bethany, 1.5 miles from Jerusalem on the east side of the Mount of Olives. He made his triumphal entry into Jerusalem on a donkey that he had mounted in the Bethphage area. After crossing the Mount of Olives, he descended into the Kidron Valley to shouts of "Hosanna"; after entering Jerusalem, he took a look around the temple area.

On Monday he entered the temple area again, and this time he drove out the moneychangers who were possibly operating in the Royal Colonnade along the southern perimeter of the Court of the Gentiles. On Tuesday Jesus once again entered the temple complex and later in the day spent time teaching his disciples on the Mount of Olives.

After resting in Bethany on Wednesday, Jesus sent "two of his disciples" (Mark 14:13) into the city to secure a room and prepare a meal so that he could celebrate the Passover with his disciples. In spite of the fact that the structure on the traditional site of the Last Supper (the Cenacle) dates from the Crusader period (at least 1,100 years after the event), it is probable that the site itself, located on the southern portion of the western ridge in a well-to-do section of town, is close to where the meal took place. Then Jesus and his disciples went down to the Garden of Gethsemane, at the western foot of the Mount of Olives, near the Kidron Valley. There, after praying for a while, he was taken prisoner.

That night he appeared before Caiaphas the high priest, Pilate the procurator, and Herod Antipas, the ruler of Galilee, who was in Jerusalem for the festival. The exact site of each interrogation is not known, but most likely the residence of Caiaphas was somewhere on the southern or eastern portion of the western ridge, and Herod Antipas was probably staying in the old Hasmonean palace on the eastern slope of the western ridge, overlooking the temple. Although Jesus may have appeared before Pilate at the Antonia Fortress, it is more probable that as ruler of the country, he was residing in Herod's palace and Jesus was interrogated, humiliated, and condemned there.

According to the gospel accounts, Jesus was led outside the city, crucified, and buried in a nearby tomb belonging to Joseph of Arimathea. In Jerusalem today two localities lay claim to these events. The first of these is Gordon's Calvary, to the north of the present-day Damascus Gate, with the nearby Garden Tomb. Although this site lies outside the ancient as well as the present-day city wall and is quite amenable to certain types of piety, there is no compelling reason to think that this is either Calvary and/or the tomb; in fact, the tomb may date back to the Iron Age (1000 – 586 BC) and thus would not have been a tomb "in which no one had yet been laid" (Luke 23:53).

More compelling is the suggestion that the Church of the Holy Sepulcher marks the spot of these dramatic events. This more traditional site was probably outside the walled city of Jesus' day and was in fact a burial ground. After his resurrection Jesus appeared to his disciples for forty days and then, on the Mount of Olives, he ascended into heaven.

▲ *Crusader entrance to the Church of the Holy Sepulcher. This church contains both the site of Calvary and the tomb of Jesus. Tradition dating back to the 4th century AD.*

During the early apostolic period (ca. AD 30 – 44) the Christian church was centered in Jerusalem. Various events took place there, such as house meetings, appearances before the Sanhedrin, and imprisonments, but it is all but impossible to pinpoint their location. In the temple precincts a paralyzed man who was sitting at the "Beautiful Gate" (probably the gate that led into the Court of Women) was healed, and the early Christians often met in "Solomon's Colonnade" (Acts 3:11; 5:12) — probably the colonnade along the inner side of the eastern enclosure wall of the temple precincts.

After Agrippa I died (AD 44), Roman procurators ruled Jerusalem directly until the outbreak of the first Jewish revolt (AD 66 – 70). During this revolt, the Romans, slowly but surely, subdued the rebels. In the spring of AD 70 the Fifth, Tenth, Twelfth, and Fourteenth legions, and their slave captives — about 80,000 men in all — advanced on Jerusalem (map p. 127).

The Jews attempted to fortify the third, or northern, wall that Agrippa I had begun, but by the end of May the Romans breached it. A few days later the second wall was breached and a siege dike was set up around the remainder of the city. The suffering within the city was severe, and in late July the Antonia Fortress was attacked and captured. From there the Romans advanced into the temple precincts, and on the ninth of Ab (August 28) the temple was burned. Titus, the Roman general who later became emperor, ordered much of the city to be razed, save the three towers just north of the Herodian palace. These he left standing as mute tribute to the greatness of the city he had just captured.

Tell Beth Shan in northern Israel. During the OT period life was focused on the tell, but during the Greco-Roman period the city greatly expanded to the area at the foot of the tell.

Why Are There Tells?

The ancients did not set out to build tells. Rather, it often took centuries for tells to develop. The following are some of the more important factors that entered into the complex process of the formation of tells.

1. People preferred to settle close to a source of fresh water — a spring, a well, or, more rarely, a flowing stream.
2. By settling on a hill or a rise near a water source, people could more easily monitor the surrounding landscape and defend themselves...
3. People preferred to live in regions with good agricultural land and/or pasturage.
4. Often people wanted to live close to major and even minor "roads," which may have led to the settlement of some sites.
5. Other sites may have developed because of their religious significance, their proximity to special natural resources, etc.
6. If stone walls, foundation walls, or even just stones from structures of previous inhabitants of a site were available, these could easily be reused in the building of a new settlement. In some areas of the country, the accumulation of mud from mud bricks also significantly contributed to the rise of a tell.

Since there was a limited number of water sources with a limited number of hills near them, and since on these sites building materials from previous settlers were often available, it was here that new settlements were built over old ones, a process that was often repeated many times over. Thus in the end, the distinctive mounds now known as tells were formed.

SCRIPTURE INDEX

SUBJECT INDEX

Salem, 40. *See also* Jerusalem

Salim, 118 – 19

Salome Alexandra, 107

Samaria, 120, 122, 78, 79, 83, 84; Alexander destroyed, 98

Samson, 60, 63

Samuel, 64, 65

Sarah, 40

Sardis, 137 – 38

satrapies, 92, 93

Saudi Arabia, 8, 9, 13

Saul, King, 64 – 68

Saul, later called Paul, 128

Sea of Galilee, 12, 115 – 17, 126

sea peoples, 62

seasons, 14 – 15; in Egypt, 18; in southern Mesopotamia, 27

Sebaste, 120

Second Intermediate Period, 42 – 43

Seleucus I, 100

Seleucid eparchies, 101

Sennacherib, 86, 87, 88

Sepphoris, 114 – 15, 125

Sergius Paulus, 129, 130

Sermon on the Mount, 116

Seti I, 45, 62

Shalmaneser III, 81

Shan, 55

Sharuhen, 42

Sheba, Queen of, 75, 71, 72

Shechem, 37, 51, 76, 78, 120

Sheep Gate, 148

Shephelah, 52, 54, 63, 66, 86

Sheshbazzar, 93, 145

Shiloh, 55, 64

Shimron, 52

Shishak, 76, 78

Sidon, 52, 89, 101, 118, 119

Sihon, 49

Simeon, territory of, 54

Sinai, 45, 46, 47

Sinai Peninsula, 20 – 21

sirocco, 15

Sixteenth, Dynasty, the, 42

Smyrna, 137

Sodom, 39, 40

Solomon: districts of, 75; monarchy of, 69 – 75; temple of, 142

Solomon's Colonnade, 149

Southern Kingdom, conflicts of, 81

Southern Levant: during the Early Bronze Age, 34, 35; during the Middle Bronze Age, 39. *See also* Levant

springs, on the slopes, 11; of Gihon, 69; of Harod, 62

Stephen, stoning of, 128

Story of Sinuhe, 41

Strato's Tower, 111

Subite, 83

Succoth, 46

Suez Canal, 46

Sumerian culture, 32

Sychar, 121

Syria, 8, 9; geography of, 23 24

Syrian Gates, 97

Taanach, 55, 70

Tabgha, 117

Table of Nations, 31

Tarsus, 97, 128, 129, 131

Tell ed-Dab´a, 45, 46,

Tell es-Sultan, 49

temperature, in Jerusalem, 14

Temple Mount, 146, 147; seized the, 125, 126

Temple of Zeus, 103, 105

temple, the: Herod's, 111; in Jerusalem, 105; location of the, 143 – 44; rebuilding of, 92; Solomon's, 75; refurbishing of, 147; Romans advanced into, 149

terrain: of Egypt, 18; of Sinai, 21; of the Middle East, 10 – 13

Thessalonica, 130 – 33

Thirteenth Dynasty, the, 42

Three Taverns, 135

Thutmose I, 44

Thutmose, III, 44, 45

Thyatira, 130 – 31, 137 – 38

Tiberias, 115 – 16

Tiglath-Pileser III, 84

Tigris River, 8, 9, 25 – 27

Timnah, 63

Tirhakah, 86

Tirzah, 78

Tobiad family tomb, 95

tomb, of Jesus, 148 – 49

tombs, Middle Bronze I, 38

topographical map, of Israel and Jordan, 12

Transjordan, 101; Mountains, 11, 12; Highway, 13

transportation, in Mesopotamia, 27; modes of, 17, 18. *See also* travel

travel, 15, 16, 17. *See also* transportation

tribal territories, 53 – 54

Tripoli, 24

Troas, 131, 133

Turkey, 23, 100

Twelfth Dynasty, 41, 42

Tyre, 89, 101, 118, 133

Tyropoeon Valley, 141

Ugarit, 24

Umm Qeis, 118

Ur, 32, 36

Uruk, 32

Valley of Elah, 66; Eshcol, 47; of Rephaim, 69; of Salt, 70

Vespasian, 125, 126

Via Egnatia, 130 – 32

Via Maris, 17

volcanic mountains, 13

Wadi el-Arish, 20

wadis, 10, 13, 17

wars, of expansion, 70

Waters of Merom, 52

Way of the Sea, 17

wheat harvest, 14

wilderness, the, 118

winds, 15

Xerxes, 92

Yarmuk River, 12, 17

yearly cycle, in ancient Egypt, 18. *See also* seasons

Zebulun, territory of, 55

Zedekiah, 89, 90

Zered Valley, 12 – 13

Zerubbabel, 93, 145

Zeus, 103, 105

ziggurats, 27

Ziklag, 67, 68

Zoar, 40